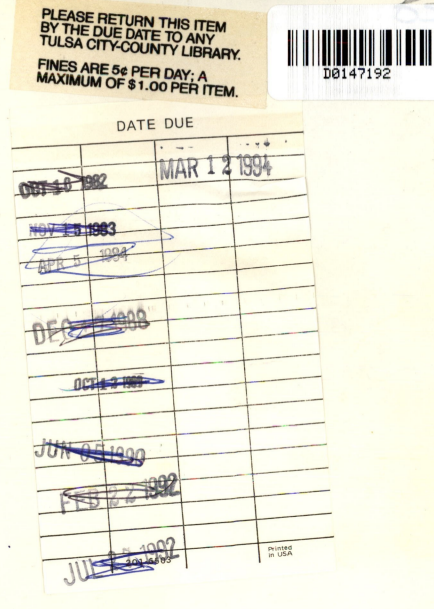

DATE DUE

		MAR 1 2 1994	
OCT 1 6 1992			
NOV 1 5 1993			
APR 5 1994			
DEC 1998			
OCT 1 2 1989			
JUN 0 5 1990			
FEB 2 7 1992			
JUL 1992			Printed in USA

301-6503

AN
ILLUSTRATED
HISTORY OF
CIVIL ENGINEERING

AN ILLUSTRATED HISTORY OF CIVIL ENGINEERING

Neil Upton

CRANE RUSSAK · NEW YORK

Published in the United States in 1976 by
Crane, Russak & Co Inc
347 Madison Avenue, New York NY 10017

ISBN 0–8448–1032–0
Library of Congress Catalogue Card Number 76–41092

Printed in Great Britain

Contents

Introduction

"The art of directing the Great Sources of Power in Nature for the use and convenience of man, as the means of production and traffic": this was the first written definition of civil engineering which was given by Thomas Telford in the Charter of Incorporation of the Institution of Civil Engineers, London, in 1828. This statement was broader than would apply today, as at that time the field included all branches of engineering not carried out by the army. Later in the nineteenth century, the Electricals and Mechanicals split off to form their own institutions.

In Medieval times the civilian engineer was a skilled craftsman or artisan, and not a professional engineer in the modern sense. Up to the eighteenth century, the only employer for the professional engineer was the military. He designed and built fortifications, roads, bridges, water supplies for camps, machines of war, guns and mines (tunnels under enemy lines packed with gunpowder). Leonardo da Vinci was so employed for a considerable part of his career. But as the complexity of civilization grew, so did the non-military application of engineering. Therefore, to distinguish this type of engineering from the military, the term civil engineering was used.

However, the civil engineer still has to contend with and combat all the unpredictable elements of Nature because he always works out of doors. Civil engineering is certainly carried out for "the use and convenience of man", and this aim is the most important part of this branch of human endeavour. Frontinus, Water Commissioner to Rome (A.D. 97–104), certainly thought so when he wrote: "Will anyone compare the idle pyramids, or those other useless though renowned works of the Greeks, with these aqueducts, with these indispensable structures?"

Civil engineering is essential to any developing society; other engineers can do little without, e.g. transport or irrigation systems. The civil engineer is a pioneer in other ways too, opening up the vast interiors of continents. The Brazilians today are doing just this in the Amazon Basin; the civil engineers go in first to clear the jungle and build roads, to dig canals and drain swamps.

In this book I have dealt mainly with civil engineering achievements in Western societies because it is here that the most significant advances have taken place and development has been most rapid. However, the erection of buildings I have kept to a minimum because I regard this as primarily the province of the architect. Telford's definition makes it clear that the major part of a civil engineer's work is concerned with transport. It still is, and bridges are the most important single part of any transport system. You will find both these factors reflected in the make up of this book. Indeed, the bridge is the crowning glory of the civil engineer; it is the purest and most beautiful construction he builds.

What of the future? There is no limit to what an engineer can do, provided there is money to pay for it. Longer bridges, taller buildings, all sorts of bigger and better projects will be built: like the Channel Tunnel under the English Channel; or perhaps a dam across the Straits of Gibraltar which would be 18 miles (29 km.)* long, bulging into the Atlantic, and 1,000 feet (305 m.) tall (the Mediterranean Sea evaporates and is fed by water from the Atlantic at a rate equal to twelve times that flowing over Niagara Falls, and this flow could drive turbines to make electricity); or like a projected tube train system along the northeastern seaboard of the United States (the train would run in a vacuum at 350 mph (560 kph), pushed from behind by atmospheric pressure). All these schemes could be built using present knowledge, they are merely on a grander scale than anything before. Perhaps new materials and techniques developed in the future will further extend the limits of civil engineering, and we may have bridges of plastic with two-mile (3 km.) spans, or dams and ship canals built by nuclear explosion.

* All measurements are shown in Imperial Measure with the equivalent in metric units. Specific quantities are converted accurately, but where the measurements are general the conversion is approximate only.

I

The Ancient Period

In the Old Stone Age, about 60,000 to 10,000 B.C., men were nomadic food gatherers and there was no conscious civil engineering. They may have made the odd bridge from a fallen tree trunk or a rope bridge from vines, but no one did such things as a full-time occupation. The society could not spare the food to carry specialists so there were no engineers. The change came with the New Stone Age about 5000 B.C., after it had been discovered that food production was more efficient than food gathering. Men became farmers instead of hunters and settled in villages and then cities. This was the Neolithic Revolution. It produced more engineering progress in 2,000 years than there had ever been in man's previous existence, because urban society with its food surplus could afford engineers and had need of them. The main contribution of the civil engineer was the control of water for irrigation and drinking, and the construction of buildings.

Water was essential for these early settlements and they appeared first of all along the Rivers Nile, Tigris and Euphrates. These rivers flood once a year—Noah's flood can be traced to the Euphrates—and the flood waters were controlled to stop damage and to be stored for later use. This required dams, storage basins, canals and levees. To judge from the mounds left in Mesopotamia in present-day Iraq, the levees to contain the rivers on a regular course must have been hundreds of miles long and at least a hundred feet wide. An Assyrian dam above Nineveh still stands 10 feet (3 m.) high, and there are the remains of a stone dyke in Syria which was $1\frac{1}{4}$ miles (2 km.) long in 1300 B.C.

There are very few remains left of the early irrigation works, but we can tell what these must have been like from later examples because once the urban society was well established around 3500 B.C. there was virtually no change in engineering techniques. One of these later examples was an aqueduct built on the orders of Sennacherib, the Assyrian king, in 703 B.C. He wanted more water supplied to his capital Nineveh so he built a 30-mile (48 km.) long feeder canal to the Khosr canalized river which carried irrigation water already for the 15 miles (24 km.) into Nineveh. At Jerwan the feeder canal had to pass over a valley on an aqueduct. This was 863 feet (263 m.) long, 68 feet (21 m.) wide, 28 feet (8·5 m.) at its highest, and solidly built with over two million stone blocks, except for five corbelled arches of 8-foot (2·4-m.) span. The picture below shows a corbelled or false arch.

A corbelled arch above the City Gate of Messene.

Immediately under the paved lining of the water channel there is a 16-inch (40-cm.) thick layer of concrete. It is proper concrete using a burnt lime cement. The Jerwan Aqueduct had been regarded as an ancient dam until it was excavated in 1933 and the arches discovered. It is therefore interesting to find that it was described as an aqueduct in a local legend. The story is that a king offered to give his daughter to the first man who could supply water to the village of Tell Kaif. One suitor started to build the aqueduct, but he died of a broken heart when he discovered that a rival had beaten him. Actually the rival only put sheets of cloth on the ground, but at a distance they looked like water.

10   The use of concrete in the Jerwan Aqueduct shows that the Romans did

not invent it as some suppose. Neither were they the first to use the true arch. It was known to the Egyptians in 3000 B.C. and used for small spans, probably constructed without the use of a supporting frame (centring) under the arch. There is an arch at Thebes still standing with a span of 13 feet (4 m.). The form of the arch was made using interlocking thin bricks, then ordinary bricks were laid on top.

Egyptian brick arches at Thebes.

The water for irrigation canals had to be raised by hand if the level in the rivers dropped. The level of the Nile in Lower Egypt was regulated by Lake Moeris in the present-day province of Faiyum, a large natural depression in the desert lying below sea level. A canal linked it to the Nile from about 2300 B.C. so that the depression could store the Nile flood. Then as the natural flow slackened it was supplemented from the lake, which was roughly circular and 50 miles (80 km.) across. Flow in and out of the lake must have been governed by the breaching or building of an earth dam. William Willcocks, designer of the first Aswan Dam (1902, page 145), put forward a plausible theory concerning the lake and the famines of Joseph. Joseph was in Lower Egypt around 1730 B.C. and Lake Moeris was in full use. Upper and Lower Egypt were separate states then and at war. The strategic value of the dam holding the lake must have been obvious to Joseph, but he was in prison with many captives from Upper Egypt where he may have heard of plans to cut the dam. He therefore told the King of Lower Egypt to store corn because famine was coming. Upper Egypt duly captured the dam and cut off the irrigation supply to Lower Egypt and there were seven years of famine followed by seven years of plenty when Lower Egypt recaptured it. It is also suggested by Willcocks that using his understanding of irrigation control, Moses crossed the "Red Sea" which was actually the now extinct Pelusiac branch of the Nile Delta which flowed to the east of the present Port Said. After the crossing of the nearly dry river bed, Moses breached the dam which was stopping the Nile water running to waste in the sea, and Pharaoh's army was drowned.

Zoser's Step Pyramid which was built about 2815 B.C. It is 204 feet (62 m.) high on a rectangular base 411 feet (125 m.) by 358 feet (109 m.).

The irrigation works of the Near East were the life blood of those early communities. They involved more engineering than all the seventy-odd pyramids put together, but it is the pyramids that are popularly connected with ancient engineering. The reason is, that the pyramids are spectacularly useless and still there. The first pyramid, built about 2815 B.C., was King Zoser's Step Pyramid at Sakkara overlooking Memphis. His engineer

Imhotep built it in six unequal stages rising to 204 feet (62 m.) from a base 411 feet (125 m.) by 358 feet (109 m.), widest on an east–west axis. Local stone was used for the inside and limestone for the facings. The entrance went down steps some distance from the north face and along a circuitous tunnel 23 feet (7 m.) below the surface to a deep shaft nearly under the apex. The last part of the tunnel has a sloping floor, so that it reaches the shaft 52 feet (16 m.) down. The shaft is 23 feet (7 m.) square and 92 feet (28 m.) deep. Zoser's tomb was at the bottom and the shaft was filled with rubble. There are other tunnels and galleries branching off the shaft 70 feet (21 m.) from the bottom. Just under the east side of the pyramid are eleven shafts 108 feet (33 m.) deep. A child's body was found in one and maybe they were tombs for members of Zoser's family.

The biggest and best of the pyramids is the Great Pyramid, one of three, five miles west of Giza. It was built for Cheops about 2700 B.C. and was 481 feet (147 m.) high on a very accurate 755-foot (230-m.) square base. Unfortunately the Great Pyramid has been a convenient source of building stone for thousands of years and now 31 feet (9·4 m.) have gone off the top along with most of its limestone facing, so it now rises in 206 steps instead of its original smooth white exterior. The entrance is in the north face 55 feet (17 m.) above the ground and a passage nearly 4 feet (1 m.) square descends at $26\frac{1}{2}°$ for 345 feet (105 m.) into the rock below the pyramid. The pole star in Cheops' day was visible from here at its lower transit. Then the passage goes horizontally for 29 feet (8·8 m.) into an unfinished chamber, $11\frac{1}{2}$ feet (3·5 m.) high and 46 feet (14 m.) long. This presumably was to be Cheops' burial chamber but then there was a change of plan. By this time the superstructure was up several courses and an ascending passage was tunnelled through the roof of the descending passage 60 feet (18 m.) from the entrance. After the ascending passage broke out at the level built to, it was constructed as part of the superstructure and its lining blocks are parallel to its gradient whereas they were not before. The passage ascends at $26\frac{1}{2}°$ for 129 feet (39 m.). It was blocked by three granite plugs which could only be placed from the inside after Cheops' burial. There is a horizontal passage from here to the abandoned and so-called Queen's Chamber which is 20 feet (6 m.) high. Carrying on the ascending passage is the Grand Gallery, 153 feet (47 m.) long, 7 feet (2·1 m.) wide, and 28 feet (8·5 m.) high to the tip of its corbelled roof. Finally there is a low narrow passage to the King's Chamber which is 34 feet (10 m.) by 17 feet (5 m.) and 19 feet (6 m.) high. The roof is made of nine granite slabs whose aggregate weight is 400 tons (400,000 kg.). On top there are five relieving chambers to relieve the roof slabs of the weight of the 300 feet (91 m.) of superincumbent masonry. Two

13

ventilating shafts run from the King's Chamber and light from the brightest star in the sky, Sirius, shines down the one in the south face when it crosses the meridian. Sirius was an important star to the Egyptians because when it rose just before the sun it heralded the Nile flood in July. From the lower end of the Grand Gallery there is a crude narrow passage to the descending passage. It may have been hewn by subsequent robbers or it may have been the escape route, unknown to Cheops, of the men who placed the granite sealing plugs in the ascending passage after his burial.

A modern engineer with modern equipment could build the Great Pyramid in two or three years with a few hundred men, but how did Cheops' engineer do it? Herodotus, the Greek historian who lived in the fifth century B.C., said it took 100,000 men working in three-month relays for twenty years, but this is probably an overestimate. The methods used to build a pyramid are not known directly. They can only be guessed at from what is known to have been available to the Egyptians, that is: the lever, the inclined plane or ramp, rollers, and unlimited time and labour. There is a suggestion in Herodotus that Cheops' unlimited resources were strained by his building effort. He sent his daughter out "to the stews" to get money. Herodotus says that each man donated a stone, with which she built the central of the three small pyramids in front of the Great Pyramid.

The first job for the engineers was to choose a site and level it. It had to be

Entrance to the Great Pyramid 55 feet (17 m.) above the ground in the north face.

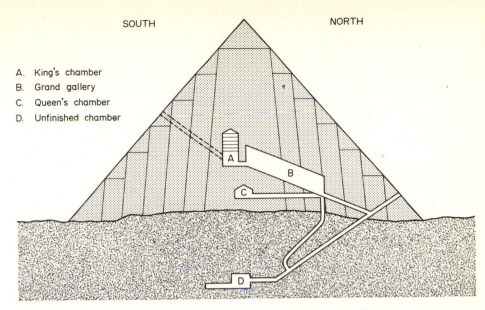

SOUTH   NORTH

A. King's chamber
B. Grand gallery
C. Queen's chamber
D. Unfinished chamber

Cross-section of the Great Pyramid looking from the east.

on the west bank of the Nile, on the side of the setting sun, and out of the flood area, but not too far away because of the water transportation needed for the stones. The surface sand was removed and the area flooded inside trenches or a retaining wall. This was a simple method of getting the base horizontal to within half an inch (13 mm.). The area was then drained before building commenced. True north could easily be found by bisecting the angle between the rising and setting of any star. All the pyramids that are in sufficient ruin show that they are made in layers, a bit like the skins of an onion as seen in cross-section. Each interfacing layer was very carefully finished, sloping at an angle of 75°. Why is not known. Perhaps the Egyptians wrongly thought it gave stability to the structure, or there may have been religious significance in the angle of the layers because it is the same as on the simple mastabas in which Pharaohs were buried before the time of the pyramids. (A mastaba was a rectangular block of masonry, small in comparison with a pyramid, but with sides sloping at 75°.) It is unlikely that a pyramid was built skin by skin because this would require very much more effort than just building upwards course by course.

The remains of the side retaining walls for sand ramps have been found and it is reasonable to assume that such ramps were used on the Great Pyramid because an inclined plane is the simplest way of raising heavy material like one of the 55-ton (55,830-kg.) roof slabs of the King's Chamber. A supply ramp would be built of sand at a moderate slope, say 15°, up to one face of the pyramid to its full width with or without retaining walls at the sides. It would have to be surfaced with wood so that it could be taken up and relaid for each course of masonry. It would also be useful to 15 have steep embankments on the other three sides to give a foothold when laying the outer casing stones. The stones would be dragged on sledges up the supply ramp and laid from the middle outwards all on one course. The passages and chambers were built with the superstructure. Mortar was not necessary to hold the heavy stones, but it was used as a lubricant and to give a good bedding when finally positioning the stones. Thus the joints of what is left of the casing average one-fiftieth of an inch (0·5 mm.) between 10-ton stones. Bosses, later removed, were left on the larger blocks to facilitate the use of levers. The pyramid would be surveyed at every few courses to make sure that the sides were going up at a steady 51° 51'. A large wooden triangle containing the correct angle and held vertical against a plumbline could have been used as an aid. This particular angle makes the height of the Great Pyramid equal to the radius of the circle whose circumference is the same as the perimeter of its square base. All true pyramids have this property which means the Egyptians were aware of the number we call "pi". When the gold-covered granite capstone had been placed, the embankments would be lowered and the casing stones dressed say 30 feet (9 m.) at a time from wooden scaffolding.

And how were the estimated 2,300,000 blocks averaging two and a half tons (2,540 km.) each quarried and transported? A contemporary account describes how Rameses IV sent a party of men to Wady Hammâmmât for some stone: the group consisted of the High Priest of Amun, Rameses-nakht as director of works, 9 senior officers, 362 subordinate officers, 10 artists and artificers, 130 quarrymen and stonecutters, 2,000 slaves, 5,000 infantry and 800 men from Ayan, a total of 8,362 which does not include the 900 who died.

The soft rocks, limestone from the north and sandstone from the south, were fairly easy to quarry. Working on a vertical quarry face up to 40 feet (12 m.) high, a groove was cut with copper chisels, and maybe picks, and the block was detached by inserting wedges. The wedges were wooden which expanded when wetted to crack the rock. Metal wedges sliding between metal plates (feathers) were also used in the normal manner. Saws "lubricated" with quartz sand could cut the blocks up if necessary. The Pharaohs' quarries were both underground and on the surface. Some underground quarries had 20-foot (6-m.) high passages running hundreds of yards into a mountain, with pillars of rock left to support the roof. The Great Pyramid is mostly made of limestone, but granite had to be used when real strength was needed. Granite is a hard rock which the Egyptians found difficult to quarry because their copper tools were softer than the rock. Small blocks, say up to 5 tons (5,000 kg.) could be cut by grooves and wedges, but the grooves would have to be made by scratching with a harder rock like flint. Copper or bronze tools could not be used for this because they would have been blunted far too quickly. Perhaps the Egyptians knew of a method of heat treating their copper to make it hard, which has since been forgotten.

Wedges could not be used on the larger blocks because uneven strains would be set up and the block would break. Their granite obelisks, for example, were separated from the parent rock solely by pounding with dolerite (a very hard rock) balls, 5–12 inches (12–30 cm.) in diameter. The surface was pounded and the resultant dust brushed away. This can be seen clearly in the quarries at Aswan, where most of the Egyptian granite came from, and in particular on an obelisk there on which work was stopped about 1500 B.C. This obelisk would have been 137 feet (42 m.) long weighing

An unfinished obelisk in its pit at Aswan. It would have weighed over a thousand tons (a million kilograms), but work was stopped around 1500 B.C. when a crack in the rock was revealed.

1,168 tons (1,186,688 kg.). The separating trench, 300 feet (91 m.) long and 2½ feet (0·76 m.) wide, was made completely by pounding. If finished it would have been 14 feet (4·26 m.) deep and then a series of galleries would have been tunnelled under, also by pounding, to be filled with packing while the remaining rock was removed from underneath. R. Engelbach, who cleared this obelisk in 1922, estimated that this work would take fifteen months at twelve hours a day using about 400 men, a third of them in the 17 trench sweeping up and guiding the pounders on poles worked by the rest of them on the surface. The obelisk could be levered up out of its pit by tree-trunk levers on both sides. The trench would be filled with sand to run under and support the obelisk as it rose. If the site permitted, the side of the pit could be removed by fire and water and wedging. A fire was built on the rock to get it hot; then it was cooled rapidly with water causing it to crack; then the slab came away easily by inserting wedges in the cracks. Next the obelisk could be rolled out on a sand bed, using levers and pulling on ropes wound round it, the ropes unwinding as the obelisk rolled. Eventually it rolled on to a sledge which was pulled lengthways, with or without rollers, to the Nile.

Rollers were not always worth the trouble. Certainly friction was reduced and less men were required to pull the stone, but the men saved here were still needed to place the rollers and to steer the object. Smaller blocks were loaded off jetties into boats up to 300 feet (90 m.) long, or on to rafts. The large blocks, like an obelisk, could have pontoon rafts built under them and they would float off when the river rose in flood. The boats and rafts were towed down the Nile to a point near the building site, and transported over land as before.

An obelisk could be erected by levering on an increasing earth bank or wooden platform, which would support the obelisk and the men working the levers. Another way could be to pull it up a sand ramp and undermine one end so that it sinks on to its pedestal. In both cases it would then be pulled to its final vertical position. Pliny (A.D. 23–79) said that Rameses tied his son to the end of an obelisk while it was being raised to encourage the workmen to be careful. It is very easy to break an obelisk by its own weight during erection because granite is a brittle material. The stone may have been strengthened by a wooden jacket at this critical time. If this Rameses in Pliny's account was Rameses II then the loss of a son may not have bothered him as he had over a hundred and several score of daughters.

Meanwhile on Salisbury Plain in southern England the Neolithic Revolution was having its effect on the ancient Britons. There they built the unique Stonehenge which, although constructed later than the Great Pyramid, is cruder because society had not reached the same stage as it had in Egypt. Stonehenge was built for religious reasons and to make observations of the sun and moon. There were three stages in its development covering about 900 years from 2200 B.C. Stonehenge 1 consisted of a circular bank 6 feet (2 m.) or more high on a diameter of 320 feet (97·5 m.), a circle of pits on a diameter of 288 feet (87·7 m.) and known as the Aubrey Holes, the so-called Heel Stone outside the bank to the northeast, and maybe a wooden structure at the centre. The material for the bank came from an

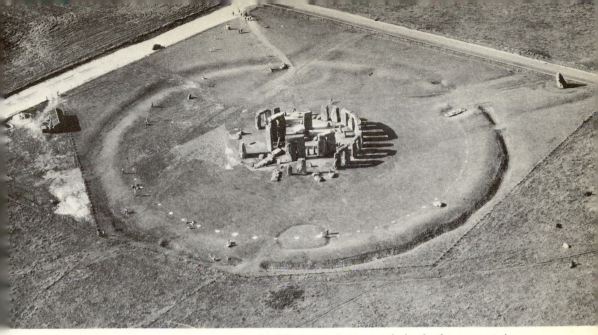

Stonehenge as it appears today from the air. There were three stages in its development covering about nine hundred years from 2200 B.C.

irregular ditch around the bank up to 7 feet (2 m.) deep, dug with neolithic tools namely deers' antlers and oxen's shoulder blades as pick and shovel. The Aubrey Holes, 24–45 inches (60–100 cm.) deep, were dug one or two hundred years after the bank and very soon refilled. Quite a few contain cremated human remains. It is possible that they were used for lunar observations and predictions by moving stones around them in turn. The Heel Stone, now leaning, is a natural 35-ton (35,560-kg.) boulder, but no doubt selected for its shape which is 20 feet (7·3 m.) long with 4 feet (1·2 m.) buried as a foundation. It does not exactly mark the point of midsummer sunrise and never has, but it could be so used and probably was.

Stonehenge 2 was built about 1700 B.C. and consisted of an unfinished double circle of 82 bluestones (mainly dolerite) at the centre of the site with an opening to the midsummer sunrise and the Avenue. The Avenue ran 2 miles (3·2 km.) round to Amesbury on the Hampshire Avon and was guarded by banks 40 feet (12 m.) apart. It must have been built as part of the ceremonies of bringing in the bluestones; and ceremony would have been reasonable if indeed the stones came from the Prescelly Mountains in Pembrokeshire which contain the nearest matchable bluestones. The biggest bluestone is 13 feet (4 m.) long and weighs 4 tons (4,064 kg.). The stones could have been dragged overland from Prescelly to the River Cleddau which runs to the sea at Milford Haven. They could have been taken along the South Wales coast by water, over the Severn Estuary, and up the Bristol Avon. From there it is possible to get to Amesbury at the end of the Avenue by various rivers with only 6 miles (10 km.) over land. Professor

Stonehenge viewed from the centre, looking through part of the Sarsen Ring at the Heel Stone in the direction of midsummer sunrise. Part of the Bluestone Circle can be seen in front of the Sarsens.

R. J. C. Atkinson has experimented with a replica of a bluestone. He found that sixty men would be needed to move a 4-ton (4,000-kg.) stone on a sledge with rollers, forty actually pulling, the rest moving rollers and steering. Pine log rafts built in two layers and about 20 feet (6 m.) square could have been used for the sea journey, but perhaps they changed to dug-out canoes lashed together for the river journey because such craft can float in shallower water than rafts.

Stonehenge 3 was what is now seen in a ruined state although the positions of the bluestones were changed a few times before their final placings. About 1600 B.C. the sarsen (sandstone) stones were brought in and erected in a 97-foot (29·5-m.) diameter post and lintel circle after the bluestones of Stonehenge 2 had been cleared away and stored. There were thirty uprights in the circle rising 13½ feet (4·1 m.) above the ground. Their average weight is 26 tons (26,000 kg.) and their cross-section is 7 feet (2 m.) by 3½ feet (1 m.). These were capped by thirty lintels held to the uprights by mortise and tenon joints, and to each other by tongue and groove joints. These are joints for wood, not stone, and suggest that building in stone was a new venture for those Bronze Age people. The lintels curved with the circle and averaged 2–3 tons (2,000–3,000 kg.) apiece and are 10½ feet (3·2 m.) long. There were six still in position when one fell with an upright on the last day of the nineteenth century. Inside this circle was a horseshoe of five trilithons open to the midsummer sunrise; a trilithon is two uprights with a lintel over. They increase in height to the central one from 20 to 24 feet (6–7 m.) plus foundations of 4–8 feet (1·2–2·4 m.), which makes the biggest stone at Stonehenge slightly less than 30 feet (9·1 m.) long and around 45 tons (45,000 kg.). The stones were dressed by pounding the surface with

The largest stone at Stonehenge which once formed part of the largest trilithon. A smaller trilithon can be seen to the left.

hard hand-held stone balls.

The sarsens were brought 24 miles (39 km.) from near Avebury in North Wiltshire. More men per ton than the lighter bluestones would be needed, say twenty-two men per ton, so 1,100 men would be needed for the 45-ton stone increasing to 1,500 on a four degree slope. One thousand five hundred men would have been the whole population for miles around and it might have taken them ten years to move the eighty-one sarsens to Salisbury Plain let alone the time needed to dress and erect them. This must have been a great event, bigger than the construction of a pyramid was to the Egyptians. The site must have been marked out and the foundations dug first because any stones in the central area would have interfered with this. The trilithons had to be erected before the circle. The foundation holes had three sides vertical and the fourth side as a slope so that when the upright was brought up to its hole on rollers and placed in position, its end was over the hole and it overbalanced on the last roller and lay on the sloping side of its hole. It could then be pulled and levered from a stack of wooden beams into the vertical position. There are no signs of earth ramps at Stonehenge so they were not used for erecting the uprights nor raising the lintels. The lintels could have been rolled up a wooden ramp by pulling on a rope wound round the stone. The lintels could have been levered up each end in turn as a stack of wooden beams was built underneath it; then at its full height it would be rolled into position on top of the uprights. The latter method is easier but more sophisticated. This society must have been fairly advanced though, to plan and socially organize such a project, let alone using the final complex to predict eclipses of the sun and moon. Finally, the old bluestones were re-erected, eventually as a repeat pattern of uprights inside the sarsen circle and horseshoe, and six other sarsens were placed upright in various positions inside the bank.

Silbury Hill, built a hundred years after Stonehenge 1, is another structure in the same area and it shows a thorough understanding of soil mechanics. It is an artificial hill 130 feet (40 m.) high, which was probably its original height. It has maintained this because there is a system of radial and circumferential walls inside the mound. The material for the hill came from the surrounding ditch, which is usually full of water in the winter.

The Egyptians' great post and lintel building was the Temple of Amon Ra at Karnak built about 1400 B.C. Some of the columns are still standing, 10 feet (3 m.) in diameter and 69 feet (21 m.) tall, and are made of large cylindrical blocks but they rest on very poor foundations. The lintel beams are around 70 tons (70,000 kg.) in weight. They were manœuvred into place up an earth ramp, meaning that the whole temple was buried in this ramp.

The Temple of Amon Ra at Karnak covered a floor area of 338 feet (103 m.) by 1,220 feet (372 m.). The columns in the centre aisle are 69 feet (21 m.) high and support 70-ton (71,120-kg.) lintels.

Post and lintel building reached its zenith with the Parthenon in Athens. Architected by Ictinus and Callicrates, it did not need to be on the grand scale because by Greek times effect could be achieved by deliberate aesthetic design, a sense of beauty and mathematical proportion. The base of a column was two or three, its height ten or twelve and the space between five or six. Thus no detailed plans were necessary and only a general plan and the ratio needed to be fixed. The columns of the Parthenon are 6 feet (1·8 m.) in diameter and 34 feet (10·4 m.) tall. They were made by placing stone drums one on top of another and held together by iron dowels sealed with molten lead. The very fine joints were probably obtained by grinding the

23

Roman engineers working with a crane powered by a treadmill. The block and tackle seen here was invented by the Greeks.

stones together. Iron was used elsewhere in the building: cramps held blocks together in the walls and the overhanging cornices were cantilevered by iron anchors. A lintel in the Erectheum was reinforced by a wrought iron bar let into a groove in its lower surface and sealed with lead. The site on the Acropolis was not big enough for the Parthenon, the new temple to Athena, and it had to jut a little over the hillside making 40 feet (12 m.) of substructure on the south end necessary. The building occupies an area 101 feet (30·8 m.) by 228 feet (69·5 m.) having eight columns at the ends and seventeen on each side.

Work started in 447 B.C. The stones were taken up to the Acropolis in carts pulled by thirty to forty oxen; some of the larger blocks were encased in a wooden drum and rolled up. Bosses were left by the masons on the stones so that they could be lifted into their positions on the building by block and tackle used in conjunction with a derrick. Then the bosses were chiselled off. The Greeks were the first to realize and use the mechanical advantage of a pulley system. Great effort was made in Greek buildings to correct the effects of optical illusion. The steps on the base of the Parthenon are convex, rising to the centre to make them look flat. The columns all slope inwards to appear vertical and individually they taper with a bulge in the middle to make them seem tall and straight. The uprights of Stonehenge are the same shape. The Parthenon was finished in 438 B.C. but decoration occupied another six years. Its roof was wooden with terracotta tiles. The Parthenon remained intact until A.D. 400 when with internal modification it became a Christian church, then a mosque in 1458. It was used as a gunpowder store in 1687 and it blew up.

Before Greek times towns were hardly big enough to warrant an elaborate water supply. It was not until the days of the Roman Empire that great aqueducts were common, but there are signs that some provision was made. For example, spigot and socket jointed terracotta pipes dating from 2000 B.C. have been found in Knossos, Crete, and a notable tunnel bringing water to Jerusalem built by King Hezekiah. In about 690 B.C. Hezekiah was under attack from the invading Assyrians led by Sennacherib. The city drew its water from a spring called Gihon which was outside the walls, so a 1,750-foot (533-m.) long tunnel was built on a curve through solid rock to the Pool of Siloam within Jerusalem. It was built from each end and the tunnellers met in the middle by listening for each other inside the hill.

As towns grew bigger more water was demanded and it had to come from farther away, the nearby spring or stream no longer being adequate. Thus a few Greek cities had aqueducts to rival the more famous Roman ones. Around 600 B.C. Eupalinus engineered the water supply to Vathy on the island of Samos. Its main feature was a 3,000-foot (914-m.) tunnel with 1,000 feet (300 m.) of mountain above. The tunnel was 6 feet (1·8 m.) wide, and as usual built from each end. The two headings missed each other by 16 feet (4·9 m.) and had to be connected by a cross-heading making a z-bend in the tunnel. The aqueduct went on into the town on masonry arcades. The city of Pergamum in the second century B.C. had an aqueduct containing water under considerable pressure, a degree of engineering sophistication which the Romans did not use. The water was carried 35 miles (56 km.) in 7-inch (18-cm.) clay pipes. Two deep valleys had to be crossed and instead of flowing across on an arcade under gravity in the Roman style, the water went down in the pipes to the floor of the valley and rose up the other side pushed by the weight of the water in the down pipes. The pressure at the bottom was 300 pounds per square inch (21 kg. per square cm.) and such an arrangement is called an inverted siphon. The pipes were anchored and their joints sealed by passing through stone blocks laid 4 feet (1·2 m.) apart in a trench.

There were scarcely any roads to speak of in the pre-Roman era apart from paved streets in towns and a few routes between towns, but only a small proportion of the latter were metalled. It is possible that some of the barrows or tumuli (artificial mounds) on the ridgeways of ancient Britain acted as directional route markers. These roads were on the ridges or crests of the land because the valleys would have been forested and swampy. Some of the prehistoric roads were so well used that they became hollow or harrow-ways. One at Willersey Hill on the border of Gloucestershire and Worcestershire is 10 feet (3 m.) deep which at two inches (5 cm.) a century means it was in use for 6,000 years.

Paved streets probably stem from Mesopotamia of around 4000 B.C. Babylon's Processional Street had a brick foundation covered in asphalt, surfaced by flagstones 3–4 feet (1 m.) square. It led to a 400-foot (122-m.) bridge over the Euphrates which had seven brick piers, 70 feet (21 m.) long and 30 feet (9 m.) wide. A notable Greek pavement crossed the 4-mile (6·4-km.) Isthmus of Corinth. It was 15 feet (4·5 m.) wide and a boat, up to a 100 tons (100,000 kg.) in weight, could be hauled along it on a cradle and rollers saving the 450-mile (724 km.) sea journey round the Peloponnesus. On the whole the Greeks had no use for roads. The city states were more or less independent and they used the sea for transport. If they needed a road in a rocky area they merely cut wheel ruts on a standard gauge of about 4 feet 8 inches (1·4 m.). The ruts were 3 or 4 inches (8–10 cm.) deep, 12 inches (30 cm.) in bumpy ground. There were passing places at intervals where the ruts were doubled.

These ancient societies, covering the periods known as the New Stone Age, the Bronze Age, and the Iron Age, obviously accomplished much in civil engineering, but the inventiveness came in their periods of growth. Once each society had reached maturity, engineering progress stopped. This may have been because the very notion of progress had not been conceived, but the stagnation came because of the universal use of slaves. The slave would have the technical knowledge to bring about technical progress, but not the incentive because his master would have taken the benefits, and the master did not have the technical knowledge because he left that to the slave. The same situation was present for most of the Roman period, too. Indeed Rome's ultimate decline can be ascribed to their lack of slaves in the later period, but this shortage led to the rise of the master-craftsman who would benefit directly from his own inventions. Thus although political power, culture and science declined in the Dark Ages after the Roman Empire, technology did not.

2

Rome

The Romans were the first proper civil engineers, even though roads, bridges, aqueducts and large buildings had been made before, but they were developers rather than originators. The Romans used their engineering consistently and deliberately throughout their empire. Such a grand manner of engineering needs money, organization, and a strong drive, and these are what Rome had with its powerful central administration. The Greeks regarded any form of manual work as beneath them. Engineering was for slaves, but in Rome it was a profession and highly respected. Many emperors took an active interest in the works they ordered and the Emperor Hadrian was an engineer himself.

Lines of communication are essential in conquering and holding any empire. This was the purpose of the famous Roman roads, only to be equalled by Napoleon for the same reason, and surpassed in the present century. No one, until the railways were built, could travel faster than say Emperor Tiberius travelling from Germany to Lyons, France, at 200 miles (322 km.) a day. In 1834 when Robert Peel was asked to be Prime Minister he was in Rome. It took him thirteen days to cover the 1,000 miles (1,600 km.) to London, and he would have travelled a lot of the way on the original road surface that the Romans had laid. Their roads lasted so long because they had proper foundations which spread the load of the road on to the well-drained sub-soil. The work was done by soldiers, in conquest because there was no one else, and in peace-time to keep them occupied.

Roman roads were direct rather than obstinately straight, but the directness is remarkable in places—in its 182 miles (293 km.) from Axminster to Lincoln, the Fosse Way was never more than 6 miles (10 km.) off the straight line. It would deviate to avoid local obstacles. The line was laid out using simple instruments and beacon fires or smoke signals. It was levelled as far as possible and ditches were dug on either side. Large stones were laid on their edges for the foundation. The next layer would be large gravel set in mortar, then small gravel to act as a bed for the pointed bases of the polygonal paving stones. The surface was cambered at either side to shed the rain and its width varied from 4 to 24 feet (1·2–7·3 m.) depending on the importance of the road. Roads not carrying heavy traffic would stop at one of the layers of gravel and have kerbstones. All this was at least 2 or 3 feet (1 m.) thick and could be 8 feet (2·5 m.) thick depending upon the site. The materials used would be decided by local conditions. The foundation was supported by piles where the subsoil could not be drained, as shown where Watling Street had to pass over swampy ground near Rochester. Here the road was founded on 4-foot (1·2-m.) oak piles linked with beams to support $3\frac{1}{2}$ feet (1 m.) of large stones, 5 inches (13 cm.) of beaten chalk, 7 inches (18 cm.) of fine beaten flint, 9 inches (23 cm.) of pebbles in black soil and finally slabs 6–8 inches (15–20 cm.) thick with fine gravel in the joints. Claudius invading Britain in A.D. 43 with elephants and four legions, needed this good road from Dover to London (later extended to Chester) to carry the invading force and their supplies. In Germany, Holland and Belgium the Romans built many log roads in the local style. This involved laying 10-foot (3-m.) log planks on bundles of brushwood. Wooden skewers were driven through to anchor them.

There were over 50,000 miles (80,000 km.) of trunk road in the Empire, that is twice the circumference of the Earth, and at least five times as many secondary roads. There were milestones on every trunk road after 123 B.C., some 8 feet (2·4 m.) high and weighing 2 tons (2,000 kg.). The oldest milestone in England was found on the Fosse Way near Leicester. Its inscription had been partly obliterated as it had been used as a garden roller for many years, but it was still possible to read that it had been erected in Hadrian's reign in A.D. 120.

The Via Appia was the first trunk road the Romans built. It got its name from Appius Claudius who was Censor of Rome in 312 B.C. It was the main road to the south of Italy, North Africa and the eastern Mediterranean, going from Rome to the ports of Brindisi and Gallipoli in the heel of Italy. Pliny thought it "a miraculous work". Trouble and expense were not spared to make this so. It was dead straight for 60 miles (96 km.) from Rome to the spa town of Terracina, crossing the malaria-ridden Pontine Marshes on a 6-foot (1·8-m.) high causeway. Wooden piles were driven into the swamp to retain the rock filling. There were other such embankments on the Via Appia, some up to 45 feet (14 m.) high and 40 feet (12 m.) wide. There were many short tunnels around 200 yards (183 m.) long, and cuttings, the biggest near Terracina running 117 feet (36 m.) deep through a marble cliff. The Romans, of course, had no explosives; they cut through rock with hammer and chisel, and fire and water.

The Via Flaminia, originally an Etruscan track, left Rome northwards up the Tiber valley, ran to the Adriatic Sea at Rimini and on to Bologna. It was continued from there as the Via Aemilia to Turin and then north through the Alps. (The important passes were the Little St Bernard, the Brenner Pass and Mont Genèvre Pass.) The road was built under the rule of Consul Gaius Flaminius in 217 B.C. but was rebuilt and greatly improved by the engineer Agrippa two hundred years later. He built the fifteen bridges on the road, some of which were blown up during World War II to stop the invading Allies. There was a 600-foot (183-m.) long viaduct over the gorge at Narni, the longest of its four arches spanning 138 feet (42 m.). The Appennine Mountains were pierced at the summit of the road by a 984-foot (300-m.) long tunnel at the Furlo Pass.

Associated with the Via Valeria, which went from Rome directly across to the Adriatic is the amazing Fucino Tunnel. Lake Fucino (now Celano) had no exit so its level rose and fell considerably. This ruined many farms and put the road in jeopardy. So an overflow tunnel was built about A.D. 40, $3\frac{1}{2}$ miles (5·6 km.) through the solid rock of Mount Salvino to the River Liris (now Garigliano). Claudius's engineer planned it 6 feet (1·8 m.) wide and 10 feet (3 m.) high. It occupied 30,000 men for eleven years, working from up to forty shafts, of which the deepest was 400 feet (122 m.). The tunnel fell into disuse when the power of Rome declined, but it was opened up again in the nineteenth century. It was the longest tunnel in the world until the Mont Cenis Tunnel was built in 1876.

Alcantara Bridge in Spain is still in use. It was built in A.D. 98 by Gaius Julius Lacer who said it would last for ever.

Appolodorus's bridge over the Danube as it appears on Trajan's column in Rome. Nothing remains of the bridge itself now.

A bridge is an integral part of a road. The optimum sites for the bridges decide the route of the road. The Romans built at least 2,000 bridges, half of them in Italy. Their biggest appears to be that at Alcantara over the River Tagus in Spain carrying the road into Portugal. Gaius Julius Lacer, Trajan's engineer who built the bridge in A.D. 98, said it would last for ever. It is still in use. Its six arches of dry stone carry 600 feet (183 m.) of road 175 feet (53 m.) above the river. The two centre arches each have a span of 118 feet (36 m.). The climate and lack of heavy traffic in Spain must be good for the survival of Roman bridges because there are quite a few there, but none left in Britain. There is a long bridge over the Rio Tormes at Salamanca and a viaduct half a mile (800 m.) long of sixty arches on the same road over the Rio Guadiana. Another famous bridge was built over the Danube by the engineer Apollodorus. There is nothing left of it now, but Trajan was evidently so pleased with it that a relief of it appears on his column in Rome. The bridge was 3,720 feet (1,134 m.) long, made of twenty-one wooden spans between masonry piers. The average span was 120 feet (36 m.) and the road was 40–50 feet (12–15 m.) wide. It is thought that Hadrian had it destroyed when he decided to hold the Empire at the Danube frontier. (With the same philosophy he built his wall from the River Tyne to the Solway Firth in northern England in A.D. 127. It is over 70 miles (113 km.) long and made of concrete faced with stone.) In some remote regions, the Romans used the "hump-back" bridge in which the road follows the line of the arch instead of being horizontal. Where a minor road crossed a shallow river a   31 ford would be built with a paved surface and guard rails.

Founding piers under water is a problem which still tests the civil engineer today. In shallow water, say up to the depth of a man, the Romans drove a circle of iron-tipped wooden piles in close formation, and excavated the river bed inside the circle under water to find a firm bottom. Then they built their foundation of hydraulic pozzuolana concrete which set under water. In the early days they used just lime for their cement but it was not strong and could crumble when dry. Around 150 B.C. they discovered that the volcanic ash from Mount Vesuvius near the town of Pozzuoli made a very strong hydraulic cement. Adding some stones, they had concrete. If the water was deeper than the height of a man, or the firm bottom too far down, they built a coffer dam of two circles of piles laced with wickerwork, and filled the annular space between them with clay. Then they baled or pumped the water out from the centre and excavated in the dry. For lesser bridges the foundations could be on piles.

Traditionally Rome was founded in 753 B.C., but it was 300 B.C. before the Romans ruled Italy and 150 B.C. before their empire was established. By A.D. 400 it was in rapid decline, so Rome was a power for at least 600 years. This could not have been achieved without the network of roads already described, but the Romans also used their engineering abilities on what Tacitus, writing about A.D. 100, called "the amenities that make vice agreeable", namely: central heating, baths, sewers, buildings and aqueducts.

The first aqueduct to supply Rome was the Aqua Appia open in 312 B.C. and the last and eleventh was the Aqua Alexandrina in A.D. 226. The biggest and best, and supplying the purest water, was Aqua Marcia built under Marcius in 145 B.C. Its source was 23 miles (37 km.) from Rome in a direct line but the aqueduct was 57 miles (92 km.) long, because, as in all Roman aqueducts, the water fell by gravity alone, and this meant that the channel had to meander along the contours to keep its steady gradient. For 50 miles (80 km.) it was underground in a covered trench except for brief moments when it had to cross a valley on a bridge, or tunnel through a ridge. On its last $6\frac{1}{2}$ miles (10 km.) into Rome it was carried on an arcade. It was the first of the high-level aqueducts and could supply water to all of Rome. Later, in 127 B.C. and 35 B.C., the arcade of Aqua Marcia also carried Aquas Tepula and Julia on their run into Rome. The three can be seen in the picture on page 33 where they are broken. The straight aqueduct in the same picture is Aqua Claudia with Aqua Anio Novus on its back. These two were completed around A.D. 50. Aqua Claudia was 43 miles (69 km.) long, the last eight on arches up to 100 feet (30 m.) high, and there are over a thousand

"The Aqueducts of Ancient Rome", a painting by Zeno Diemer. Five aqueducts are seen here, riding on two arcades.

spans of 18–20 feet (5·5–6 m.). This required 560,000 tons (560 million kg.) of quarried stone and fourteen years to complete. The Emperor Domitian (A.D. 81–96) shortened the Aqua Claudia by tunnelling 3 miles (4·8 km.) under Monte Affliano. He also took some of its water via an inverted siphon (pressure pipeline) to his palace on Palatine Hill. The pipe, made of lead 4 inches (10 cm.) in diameter, dropped 133 feet (40·5 m.) to the bottom of a valley before rising again. The pressure at the bottom would be 60 pounds per square inch (4·2 kg. per square cm.). The Romans did not use inverted siphons much because they had a limited capacity.

The worst water came from Aqua Alsietina. "Positively unwholesome," said Frontinus. Sextus Julius Frontinus, former Governor of Britain, was appointed by Emperor Nerva in A.D. 97 to sort out the chaos of the then nine aqueducts into Rome. They were in a bad state and water was being stolen all along the route and in the city. Frontinus put a stop to the corruption and restored the 350 miles (563 km.) of aqueducts in his care. Augustus built the 20-mile (32-km.) long Aqua Alsietina to supply water in 5 B.C. for his Naumachia Lake Stadium where sham naval battles were fought. The excess water was used in the city.

The amount of water supplied to ancient Rome per head of population   33 was similar to that of any modern city today, but there the comparison ends because the water was used differently. The Roman water was running constantly—it is not possible to turn off a non-pressurized aqueduct. It ran into 592 public fountains (A.D. 97) and six public baths. Any excess cleaned the streets and flushed the sewers.

The remains of well over two hundred Roman aqueducts have been found and at least forty cities had a system in the same class as Rome's. The Pont du Gard in southwest France is well known. It carried the 25-mile (40-km.) aqueduct supplying water to Nîmes 160 feet (49 m.) above the Bornègre Ravine. The lower spans are 50–80 feet (15–24 m.) and the thirty-five on top are 12 feet (3·66 m.) each. The engineer was Agrippa who built most of Nîmes. The arcade in Segovia was built in Trajan's time, restored in the fifteenth century, and still carries water. It is 2,700 feet (823 m.) long and 135 feet (41 m.) high.

All these road bridges and aqueduct arcades incorporated the semi-circular arch; the Romans used no other shape. The arch enabled them to bridge much bigger gaps than ever before and this logically went into their buildings. If an arch is elongated at right angles to its span and used as a roof, it is called a vault. The Basilica of Constantine in Rome which had a

The Pont du Gard carried water to the town of Nîmes in southwest France. It is 160 feet (49 m.) high.

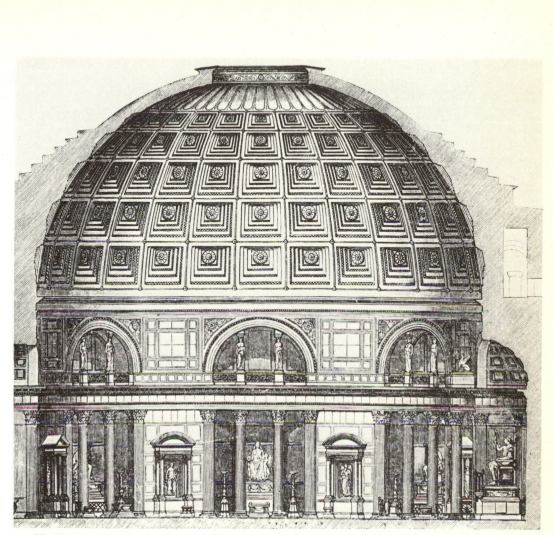

The biggest dome that the Romans built was the 142-foot (43-m.) diameter Pantheon in Rome.

vaulted nave 83 feet (25·3 m.) wide was typical of many throughout the Empire. A dome is really a spherical arch and it can be seen in the 142-foot (43-m.) diameter of the Pantheon in Rome which is still standing. The earlier vaults and domes were made of radially laid wedge-shaped or plane blocks, but in the first century A.D. the technique of laying concrete on a temporary wooden framework was developed and the ribbed dome of the Pantheon was so built.

3

Medieval-Style Water and Navigation Works

Water is essential to any community, of course, but it was especially so to western Europe in the period before the industrial revolution. Not only was it their life blood for drinking and irrigation, but it also supplied the power and the transport for their various industries. The sea was an integral part of the transport system, both for internal and foreign trade, particularly for Great Britain. Britain, in fact, was late in building canals because she was well endowed with a continuous coastline. France was therefore the first country to invest in an extensive canal system and Britain was early in the field of lighthouse development.

The ancients had built artificial waterways primarily to supply water to crops and, later, to towns. Occasionally they were used for transport. In medieval times some canals were built purposely for transport. Charlemagne appears to have been the first in Europe to plan a grand canal: it was to link the Rivers Danube and Rhine. No one had any idea of the respective levels or anything else. Work began in A.D. 793 with the digging of a channel 300 feet (91 m.) wide, but was stopped after a mile as rain, swamp and quicksand closed in. A less ambitious affair, and therefore successful, was the Naviglio Grande taking water and boats across 30 miles (48 km.) of plain to Milan. It was finished in 1258 and drew its water from Lake Maggiore.

Rivers were made more navigable by putting weirs across them to hold back the water and keep the river deeper for bigger boats. They also gave a head of water for a water wheel. The boats got through the weir via a flash lock. This only had one gate and therefore led to a conflict of interests between the boatmen and the millowner because the latter would lose his water power when the gate was opened. At Exeter in southwest England this was resolved eventually by building a canal parallel to the river. During the fifteenth century the Earls of Devon had built several weirs on the River Exe but they were reluctant to open them for navigation, thus cutting Exeter off from the sea. The city took legal action, but even with an Act of Parliament to restore navigation, they could not move the Earl of Devon, so they

Rimmer and paddle flash lock on the River Thames. The flash lock has only one gate, as opposed to the modern pound lock which has two. In this particular type of flash lock, the paddles fit together in a frame, and are withdrawn or replaced one by one.

hired John Trew in 1563 to make them a ship canal, which he did on the west side of the river. He put another weir across the river to provide water for the 10-foot (3-m.) deep canal. It was made wider and deeper in 1819 and is still in use. John Trew used a pound lock to take the ships between the different levels of the canal and sea.

A pound lock is the lock in use today having two gates enclosing a 'pound', instead of the single gate of the flask lock. Its origin is obscure. In the thirteenth century the city of Bruges was the port for northern Europe. It was linked to the North Sea by a low-lying canal which was often swept by storms. A dyke (embankment) was built across in the north to protect it from the sea. The canal went through the dyke via two gates at each end of a wooden enclosure. This was a prototype pound lock and it existed in 1234. Leonardo da Vinci used six definite pound locks when he linked another canal in Milan to the Naviglio Grande in 1497. The difference in levels was 34 feet (10·4 m.).

The next important step in canal work was taken in France with the Briare Canal which took from 1604 to 1642 to build. It was the first canal in the western world to cross a watershed, in this case the ridge of higher land between the Loire and Seine river systems. It used forty-one locks over its 34 miles (55 km.) and paved the way for the bigger Canal du Midi, also known as the Languedoc Canal. François Andréossi, who spent some time in Italy, had a hand in the planning and building of both canals, but the chief engineer and promoter of the Canal du Midi was Pierre Paul de Riquet. There were 119 locks in its 148 miles (238 km.) and a 500-foot (152-m.) long tunnel at Malpas. This was the first tunnel in the world which used gunpowder extensively for blasting. The canal left the River Garonne at Toulouse, rose 207 feet (63 m.) in the first 24 miles (39 km.) via twenty-six locks to its summit level near Carcassonne, and then gradually fell 620 feet (189 m.) to the Mediterranean Sea at Sète. Several thousand men and six hundred women were employed in the fourteen years it took to build. The channel was made 6 feet (1·8 m.) deep and 144 feet (44 m.) wide. Later it was extended down the Garonne to Bordeaux, thus linking the Mediterranean to the Atlantic. Water was supplied to the summit level from a reservoir at St Feriol, 7,200 feet (2,195 m.) long, 3,600 feet (1,097 m.) wide and 132 feet (40 m.) deep, held back by an earth dam. De Riquet died in 1680, the year before the canal was opened by Louis XIV.

Most early medieval towns got their water directly from wells and springs, and some of the wells were deep: Nuremburg Castle had a well 335 feet (102 m.) deep which at the beginning of the thirteenth century, had taken prisoners thirty years to hew through solid rock. A few Roman aqueducts had been maintained and some others rebuilt, but as the medieval towns grew these supplies were no longer adequate and extra water was brought in by canals specially constructed for the purpose.

In 1613 Hugh Myddelton, a wealthy goldsmith, had spent his fortune on an aqueduct to bring another thirteen million gallons of pure water per day to London. The main source was 20 miles (32 km.) away at Chadwell in Hertfordshire, but the actual water course was 39 miles (63 km.) long because it followed the contour line of the country falling two inches per mile to its terminal reservoir at Islington. The "New River" got as far as Enfield and then the Company's money ran out. The City Corporation would not help but James I put up half the money needed in return for half the profits and so it was completed after four years' work. The open channel was 10 feet (3 m.) wide with an average depth of 4 feet (1·2 m.), and there were forty sluices to control the water. It required more than 200 wooden bridges and in some places there were 8-foot (2·4-m.) high embankments. At Bush Hill near Edmonton the water was conveyed in a lead-lined wooden trough which was 660 feet (201 m.) long and rested on 2 foot 6 inch (76 cm.) brick piers. Another over a valley approaching Islington was 460 feet (140 m.) long and the piers 17 feet (5 m.) high. These timber structures were replaced in the eighteenth century by clay embankments.

The water was raised from the circular reservoir at Islington to a high-level basin by "a great engine worked by six sails and many horses" according to Daniel Defoe, in order to supply the higher parts of the Metropolis. The mains pipes of the New River Company were made of elm bored from tree trunks with the bark left on. Y and T junctions were made by finding part of a tree with these shapes. The pipes varied from 2 to 10 inches (5–25 cm.) bore and 6 to 10 feet (1·8–3 m.) long, and there were probably up to 400 miles (644 km.) of them. The service pipes to houses were lead. The Company stayed in existence until 1904 when the Metropolitan Water Board took them over, but the New River still supplies London with some of its water.

In 1590, with the business of the Armada over, Francis Drake promoted the building of a similar 17-mile (27-km.) canal to bring water to Plymouth from the River Meavey 12 miles (19 km.) away on Dartmoor. It was 6 or 7 feet (2 m.) wide and only took a year to dig because for half its length it used an older canal. Drake was granted a concession to install six water mills on the waterway. The owners of existing mills on the River Meavy did not like this so they appealed to Parliament in 1592 but to no avail. Francis Drake, himself M.P. for Bude in Cornwall, was chairman of the committee who considered the appeal.

However, at Plymouth the real civil engineering feats were battles against water, namely the sea, rather than for water as in the works described above. Plymouth harbour itself had always been naturally deep and safe, but Plymouth Sound was open to the sea and ships anchored there were often lost. In 1806 the Admiralty asked John Rennie to build sea defences. He planned a breakwater of 2–10-ton (2,000–10,000 kg.) blocks 3,000 feet (914 m.) long with additional endpieces 1,050 feet (320 m.) long sloping in at 20°. Work started in 1811 with the tipping of the stone blocks and by March 1813 they were above the low water level. The work was still proceeding at the rate of 1,030 tons (1 million kg.) per day when stormy weather came in January 1817 and rough seas threw 5-ton (5,000-kg.) stones from the seaward to the harbour side. Rennie's sons, John and George, carried on after their father died in 1821 and the breakwater was eventually finished in 1848. Its surface was faced with fitted masonry blocks. The 3,670,444 tons (3,730 million kg.) of large heavy rubble had been dumped to make the breakwater which still stands to this day 20 feet (6 m.) above the low water mark.

A more spectacular struggle with the sea was on the triple Eddystone Reef, 14 miles (23 km.) out from Plymouth on the approach to the Sound. This isolated underwater mass of rock is 600 yards (550 m.) long. The centre ridge, roughly north–south, is exposed for 200 yards (183 m.) except at high tide. The other two ridges, always covered, splay out from the north end 50–100 yards (45–90 m.) away. Due to their shape and position the sea is nearly always rough. The toll on shipping was getting embarrassing for Plymouth and a man was sought who could build a lighthouse.

Henry Winstanley, inventor and eccentric, businessman and showman, had a house in Suffolk open to the public. It was full of mechanical gadgets and practical jokes. There was a chair which locked its arms round the sitter, and another which descended 10 feet (3 m.). In a house in Piccadilly, London he had a similar set-up incorporating water-powered waxworks. He also owned five ships, two of which were lost on the Eddystone reefs. He was the man chosen to build the Eddystone lighthouse.

The exposed rock was 30 feet (9 m.) wide, sloping at 30° to the west and dropping vertically to the east. Winstanley started in the summer of 1696, and in the four months of that year's working season he and his men managed to drill twelve holes in the rock and fit iron stanchions in each, sealed in with molten lead. It took on average eight hours to get to the rock and work was only possible for one or two hours at low water. The rock was very hard and the holes were made by hand with picks. In the second season, July to October 1697, a solid stone masonry tower was built, 12 feet (3·66 m.) high and 14 feet (4·26 m.) in diameter. The plan called for a solid base, then a hollow portion and a wooden top section. England went to war with France in this year and Winstanley was captured from the rock by a French privateer. Louis XIV set him free at once, saying that France was at war with England, not humanity. During the third season the base was enlarged to 18 feet (5·48 m.) high by 16 feet (4·83 m.) in diameter and enough of the upper section completed to enable candles in the lantern to be lit on 14 November 1698. The work proceeded quicker in the third season because men could live on the rock in the rooms of the lighthouse and the materials could be stored there, too. There were celebrations in Plymouth when the lantern was lit, but Winstanley and his men missed them because they were stranded in the lighthouse by stormy weather for five weeks.

After the first winter of use it became apparent that the lighthouse was not big enough: the waves could reach the lantern and caused the whole tower to shudder. In the summer of 1699 the tower was encased in another one, bringing the diameter up to 24 feet (7·3 m.), and the overall height was increased from 80 to 120 feet (24·4–36·6 m.). Bands of iron were placed around the cement joints to stop them eroding. The lantern was an octagonal room 15 feet (4·6 m.) tall with sixty candles in a hanging basket.

The night of 26 November 1703 was the only known time that Britain has ever been struck by a hurricane. It was centred on Liverpool and all the

Winstanley's second lighthouse was built in 1699. He was killed when it was destroyed in a storm in 1703.

south of England was devastated. Defoe touring a part of Kent counted 17,000 trees uprooted before he gave up. One hundred and fifty ships were lost and 8,000 sailors are reckoned to have died. On land 123 people were recorded as officially killed by the hurricane; there must have been many more. Winstanley was in his lighthouse that night seeing to some repairs. In the morning there was nothing left but a few broken stumps of iron sticking out of the Eddystone rock. Two nights later, the *Winchelsea*, bound for Plymouth from Virginia laden with tobacco, wrecked herself on the reef. She was the first ship to do so for five years.

The replacement lighthouse, which took from 1706 to 1709 to build, was the work of John Rudyerd. He designed it on the principles of a ship. It was made of wood to give some flexibility under stress, and its shape was slender and smooth so that it offered very little resistance to the waves washing round it. Horizontal steps were cut on the sloping foundation rock with holes for thirty-six iron keys. These quarter-ton (250-kg.) irons were dovetailed into their holes and sealed in with molten pewter. Their purpose was to anchor the heavy timber and masonry base to the rock. The base had a diameter of 20 feet (6 m.) at the bottom and was 36 feet (11 m.) high. It was solid except for a 7-foot (2·1-m.) square shaft taking a spiral staircase from the entrance half-way up to the first of four floors. Seventy-one oaken uprights, 9 inches (23 cm.) thick, encased the base and carried on upwards for a further 34 feet (10·3 m.) to act as the walls for the living quarters, and support the lantern on the top. A mast ran up the centre as a backbone and all the wood was covered in pitch. It was a good lighthouse but it was destroyed by fire on 2 December 1755 in its forty-sixth winter.

One of the keepers, Henry Hall aged ninety-four, swallowed some molten lead from the roof of the lantern, while throwing a bucket of water upwards at the fire. He lived for twelve days, eating and drinking normally, before he died. A post mortem yielded a 7-ounce (200-gm.) lump of lead in his stomach, and this macabre casting is now in the Royal Scottish Museum, Edinburgh.

The job of building a new Eddystone lighthouse was taken on by John Smeaton (1724–92), the first man to call himself a "civil" engineer. This term did not have the narrower meaning it has today, but was used to cover all aspects of engineering as opposed to military engineering, Smeaton's forte, in fact, was as a mechanical engineer. Smeaton decided against Rudyerd's use of wood and returned to stone for the new lighthouse. He planned a shape with curving sides rising from a broad base, "Like a tree trunk," he said. The stone blocks weighing up to 3 tons (3,000 kg.), were to fit together in a three-dimensional jigsaw of dovetails, with the courses

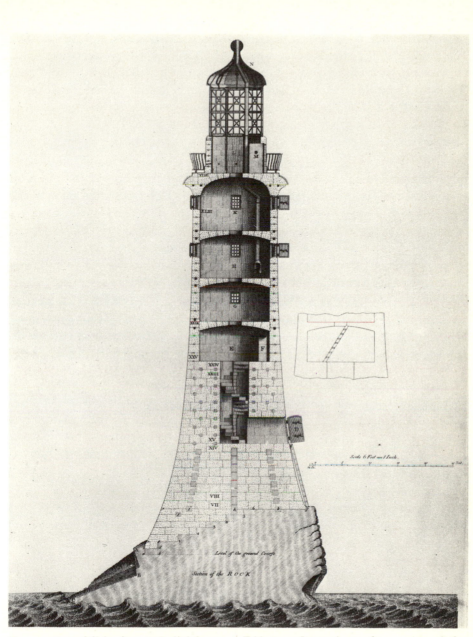

Cross-section of John Smeaton's lighthouse at Eddystone. It was in use for 123 years before it was replaced by the present lighthouse.

locked together by wooden wedges, marble blocks and wooden dowels. Ultimately, 1,493 stones, total weight 988 tons (1 million kg.), were laid, each one prefabricated and numbered to fit into its individual place. There were four circular rooms of 12-foot (3·66-m.) inside diameter and walls 26 inches thick. The rooms had domed roofs whose side thrust was taken by chains sealed in a groove round the tower at each level.

Smeaton arrived in Plymouth on 27 March 1756. The 200-mile (300-km.) journey from London had taken him six days. In the next two months he was only able to go out to the rock ten times and then only able to land on four occasions. He discovered that Winstanley's lighthouse had gone in one piece, taking a lump of the rock with it, and the only trace of the Rudyerd lighthouse was the foundation steps. Most of that season was spent in making the detailed design, organizing workyards for the stonemasons and, most important of all, the fitting out of a 50-ton (50,000 kg.) sloop as a floating workshop and dormitory. This meant that his men were permanently close to the rock and could take advantage of every half hour of calm weather for working. There were two gangs of twelve men working alternate weeks in the yards and at the rock. On 12 June 1757 the first stone was laid on the lowest of the foundation steps. It weighed $2\frac{1}{4}$ tons (2,286 kg.). By the end of that season, early October, nine courses of solid masonry had been built to 12 feet (3·66 m.) above the bottom of the rock. Work started again in July 1758 taking the tower up to the first room. The lighthouse was finished in the next working season and the light was lit on 16 October 1759. It came from twenty-four six-pound candles. The height of the lighthouse was 85 feet (25·9 m.) to the gilded ball at the top. No one had been killed or injured during the construction. It had been a straightforward job, but carried out in extremely difficult conditions, which called for a high degree of organization. From 1860–95, Smeaton's Eddystone lighthouse appeared with Britannia on the British penny. It reappeared in 1937 and stayed until decimalization.

The present lighthouse was built by James Douglass from 1878–82 because Smeaton's had become obsolete and the sea had seriously undermined the rock on which it stood. Douglass's lighthouse is twice as tall and used four times as much stone. It was built 40 yards (37 m.) away using all the tools of modern civil engineering; that is, pneumatic drills, steam winches and cranes, quick-setting Portland cement. Douglass also had the use of a twin-screw steamer capable of a speed of 10 knots and of carrying 120 tons (120,000 kg.) of stones. The top portion of Smeaton's tower was dismantled and re-erected on Plymouth Hoe as a tribute to the first "civil" engineer.

4

The Masonry Arch

Apart from the water-works described in the last chapter, the arch was the only other significant field in which progress was made by the medieval civil engineer. The principle of the arch was known to the Bronze Age societies, but it was not used on a grand scale until the Roman engineers built their great semi-circular spans.

Stone is strong in compression but weak in tension. The arch is therefore the best way of using such a material in spans of over 15 feet (5 m.) because in an arch the stresses are all compressive. A masonry arch is made of wedge-shaped stones (voussoirs) which all lean on one another and lock themselves together by friction. Thus an arch will not support itself until all the voussoirs are in position so they have to be held up, usually on wooden centring, until the final keystone is placed. A roadway can be built directly on the arch to produce a hump-back bridge, or the sides of the arch (spandrels) may be filled in to support a horizontal road.

Bridge building went into a decline in the Dark Ages but the skill was recovered in the Middle Ages. Around the twelfth century several monastic orders appear to have sprung up all over Europe called Bridge Brothers, whose aim was the well-being of travellers. There is no hard evidence of their official existence, only legend, but certainly most of the early medieval bridges were engineered by monks or priests. The most famous of them was the Pont d'Avignon built by St Bénezet.

The story is that Bénezet, a young shepherd, interrupted a church service in Avignon one day to say that God had told him to build a bridge over the River Rhône. To prove his point and mark the spot, he picked up a huge boulder and carried it to the river bank. No man alone could have lifted this stone so the towns people quickly raised the money for the bridge. Actually Bénezet had already built a bridge over the River Durance at Malpas in southeastern France in 1167, which would indicate that this was no shepherd lad but an accomplished engineer, and as such, would have no difficulty rigging up a lever or block and tackle to single-handedly shift a

The four remaining arches of the Pont d'Avignon built between 1177–87.

large rock. Work began on the bridge at Avignon in 1177. The Rhône here has two branches around the island of Barthélasse, and the bridge probably had eight spans over each stream and five on the island making it about 3,000 feet (914 m.) long. In 1385 Pope Boniface IX had one of the arches destroyed as a defensive measure for the town. The bridge was repaired by Clement VI, but it was cut again during the siege of 1395 and not restored until 1418. In 1602 four arches fell, two more collapsed in 1633, and ice broke all but four in 1670. The four remaining arches have spans varying from 101 to 110 feet (31 to 33 m.) on piers 25 feet (7·6 m.) wide, with triangular cutwaters up and down stream. The Romans only put cutwaters on the upstream end of a pier but Bénezet had realized that they were just as important downstream to reduce the eddy currents' erosion of the foundations. The arches are parts of ellipses with the major axis vertical, that is, rounded pointed arches. Bénezet died in 1184 and was buried in a chapel on one of the piers before the bridge was finished in 1187. It is 16 feet (14·9 m.) wide overall and carried a 12-foot (3·7 m.) road. St Bénezet is the only engineer whose labours have been considered worth canonization.

More representative in its engineering than the Pont d'Avignon was Old London Bridge. The typical medieval bridge was clumsy in appearance but sturdy. Ignorance of the stresses involved led to the use of excess material and many short spans, usually between 15 and 75 feet (4 and 23 m.). The pointed arch was popular because it was easier to build, and its high rise, relative to its span, reduced the side thrust that is always present in any arch. The piers were massively built to take this side thrust, often blocking well over half the stream. The stonework was never so precise or fine as the Roman's, and mostly consisted of a veneer of masonry filled with rubble. The lime mortar used was ineffective, taking years to set properly. In wet conditions the stones were set in wax, or pitch, or hot resin. Wooden or cloth patterns were used at the quarry to cut the stones to size. The piers were not evenly spaced, but sited where the foundations would be easiest to lay. Foundations could be made merely by dumping large stones or baskets of gravel until a mound was formed on which masonry could be built.

Not many river beds lent themselves to this crude method of construction and most medieval bridge piers were founded on wooden piles, as permanently wet wood does not rot. For example, Old Rochester Bridge over the Medway in England, finished in 1393, had 10,000 wooden piles which were iron shod and 20 feet (6 m.) long. Their heads were covered with elm planks below the low water level and masonry built on them. The pile-driver consisted of a derrick with a pulley on top, and an iron or wooden weight to be hauled up by hand and dropped on the pile. When a new bridge at Rochester was being built in 1458, eighteen men used two pounds of tallow for lubricating their pile-driver, each working for fifty-eight tides at three-pence per tide.

There has probably been a London Bridge for over 2,000 years. There certainly was in Roman times when it was made of wood. A wooden bridge is in constant need of repair so it may have disintegrated between the Romans' withdrawal and the Saxon invasion. There are many Saxon references to a bridge, including a description of the drowning of a witch in A.D. 984. A series of natural disasters damaged it in the eleventh century: fires, flood and ice, and it was carried away by a storm around 1090. Fire damage again in 1136 caused it to be completely rebuilt of elm wood by Peter, who was chaplain for St Mary's, Colechurch in the Poultrey in the city of London, and the Bridge Master. Another catastrophe must have struck soon after because Peter decided in 1176 to build a bridge of stone.

The site was to the west of the elm bridge, and the new bridge was made up of nineteen unequal, pointed arches supported by nineteen piers which varied from 17 to 26 feet (5 to 8 m.) in width. The spans, twenty in all, ranged from 15 to 34 feet (5 to 10 m.); one, over the seventh gap from the south, was a wooden drawbridge for the defence of the city from the south, and to allow the passage of ships. The eleventh pier from the south was bigger than the rest—it projected 65 feet (20 m.) beyond the general line to the east—and on it was built a chapel. Peter Colechurch was buried there when he died in 1205, four years before the bridge was finished.

The bridge was probably built from each end, a pier and an arch being completed on average every eighteen months. The foundation piles, 6 or 7 feet (2 m.) long and 10 inches (25 cm.) square, were chiefly of elm and driven from barges to the low water level in three rows round the circumference of a pier. The inside was filled with loose rubble, and oak sleepers, 9 inches (23 cm.) thick, were laid over this. All the work seems to have been done inside a larger enclosure of piles, later filled in with rubble to make the "starlings" which protected the foundations from erosion by the tide. The external walls of a pier were built directly on the oak sleepers and the stones held together with iron cramps fixed in their sockets in the stones by molten lead. The lower joints of the masonry were set in a cement of pitch and resin. These walls were filled with loose stones over which a liquid lime cement was run. Then the wooden centring for the arches was erected between two completed piers and the two layers of voussoirs laid in the form of a pointed barrel vault. Next the spandrels were built between the arches and over the piers, and filled with more rubble to support the gravel roadway.

Detail of Old London Bridge in the seventeenth century showing the "starlings" protecting each of the piers, but also causing an obstruction to the flow of the river.

The bridge was 906 feet (276 m.) long, but the massive piers reduced the width of the stream to 503 feet (153 m.). This was further reduced by the "starlings" to 245 feet (75 m.) at half tide, and when the water wheels were put under some arches, to 160 feet (49 m.). This was one-sixth of the original width of the stream. Old London Bridge was in effect a dam with holes in it which produced a 5-foot (1·5-m.) maximum difference in the water levels on each side. This was a great hindrance, and even danger, to boats; a point made by the boatmen in 1756 when objecting to the rise in tolls for passing under the bridge. They reckoned that about fifty of their number were killed by the current under the bridge every year. Another effect of this constriction of the Thames was to hold back the tides and make the water above the bridge much more still. This made it easier for the river to freeze over and on many occasions the ice was thick enough to support ice fairs and even the roasting of oxen. After the severe winter of 1281, the pressure of the ice floes, jammed and building up, carried away five arches and four piers. At the time the bridge was in a poor state of maintenance because Henry III's wife Eleanor had spent the Bridge Trust monies on herself. There were houses on Old London Bridge from the start until they were all pulled down for the 1762 widening. The buildings were on either side of a 12-foot (3·66-m.) roadway, so only 4 feet (1·2 m.) of the fronts were resting directly on the bridge, the rest of the buildings hung out over the river supported by struts and cantilevered beams. In 1482, a block of houses just fell into the river. The situation of the houses was even more precarious after the Great Fire in 1666 when some were rebuilt with the road almost at its full width of 20 feet (6·09 m.).

Demolition of Old London Bridge 1832, showing the original arches flanked by those added in 1762.

West side of Old London Bridge after the houses had been removed and the bridge widened in 1762.

By 1750 the bridge was inadequate for its traffic so George Dance, the City Surveyor, was instructed to prepare plans to widen the existing structure. His scheme of demolishing all the houses, extending the arches by 13 feet (3·96 m.) on either side, and making two centre arches into one, was approved in 1756 and work started on it in the following year. The traffic was carried meanwhile on a temporary wooden bridge erected on the west ends of the starlings. Arson was suspected when it burned down on the night of 11 April 1758 and Parliament ordered the death penalty without clergy for any saboteurs. The east side of the widening was finished in 1761 and the west side in 1762. The new Great Arch in the centre was now the easiest way for the river through the bridge. This produced a powerful scouring current, and the river was 3 feet (1 m.) deeper within three months and threatening to undermine the foundations. John Smeaton recommended that the stones from the recently demolished City gates should be laid on the river bed under the Great Arch. This was done in 1760 but the scour from the Great Arch continued to give trouble. By 1800 it was clear that the old bridge had reached the end of its useful life. It was a hindrance, too, to the increasing river traffic.

Designs for a new London Bridge were sought and over thirty received, including one from Thomas Telford. This was for a single 600-foot (183-m.) cast iron span rising 65 feet (19·8 m.) above high water, the 6,500 tons (6½

Telford's design for a new London Bridge, a single cast iron arch of 600 foot (183 m.) span.

million kg.) of iron carrying a 45-foot (13·7-m.) wide road. Telford's bridge was technically feasible, but as it was at a high level, it would have needed approach roads on ramps and to acquire the land for these was too expensive. A great pity, because it was more symbolic of the age than John Rennie's backward looking masonry arches, which was the design carried out. In fact Rennie died in 1821 and it was his son who actually built the bridge. It had five semi-elliptical arches, whose spans were 130, 140, 152, 140 and 130 feet (39·6, 42·6, 46·3, 42·6 and 39·6 m.), used 120,000 tons (120 million kg.) of stone, and was 56 feet (17 m.) wide.

Rennie's New London Bridge. A view from the north showing the final stages of the demolition of the old bridge—three piers can be seen down to the water level and an arch on the left. Rennie's bridge had five elliptical arches whose spans ranged from 130 to 152 feet (40 to 46 m.).

John Rennie (1761–1821) was the son of a farmer in East Lothian and worked for a local millwright, using the money he earned to pay for his education at nearby Edinburgh University where he graduated in 1783. Within two years he had built his first bridge: a modest affair on the Edinburgh–Glasgow Turnpike which had three semi-elliptical arches. He designed all the machinery for the Albion Flour Mills, London, which in 1788 was the first mill in the world to be powered by steam. This led him to set up in London as a civil engineer. His list of works is impressive but none of them was revolutionary; he used traditional methods and traditional designs. One exception may have been Old Southwark Bridge, completed in 1819 after six years' work. He used cast iron for its three arches, and at 240 feet (73·15 m.) its centre span was the biggest iron span in the world at that time. But even with this new material his design was traditional. He merely substituted iron boxes, 6–8 feet (1·8–2·4 m.) deep and 13 feet (3·96 m.) long, for the usual stone voussoirs. Old Southwark Bridge was replaced in 1921.

The first pile for the foundations of the New London Bridge was driven on 15 March 1824. A timber raft was laid over the pile heads and on this the masonry piers were built inside coffer dams (page 32); exactly the same as 623 years previously, except that steam engines pumped the coffer dams dry. On 15 June 1825 the first stone, a 4-ton (4,064 kg.) block of Aberdeen granite, was laid with ceremony in the south pier. There were seats for 2,000 people inside the coffer dam and standing room for 400 more on its floor 45 feet (13·7 m.) below high water. It was covered by a marquee, and a wooden pile bridge had been built to it from the old bridge 100 feet (30 m.) away to the east. The Duke of York was there and the combined bands of the Horse Guards and the Royal Artillery, but the stone was actually laid by the Lord Mayor of London.

The wooden centring for Old Waterloo Bridge, opened in 1817.

The arches were built using the same methods as Rennie had devised for his Waterloo Bridge, opened in 1817. There the 50-ton (50,000-kg.) centrings had been built on the shore as complete units. They were floated out between the piers and lifted into position by four screws, 8 inches (20 cm.) in diameter, and fixed to cast iron boxes in the barges. Usually centrings were built piece by piece *in situ* after the piers were finished. The keystones were driven home with a heavy wooden ram, hard enough to lift the arches off their centrings. Old Waterloo Bridge had nine semi-elliptical arches, each of 120 feet (36·6 m.), which rose 30 feet (9 m.) above high water, and its piers were 20 feet (6·1 m.) wide. The present Waterloo Bridge was opened in 1942.

Rennie's London Bridge was opened by William IV on 1 August 1831. Over forty men out of the 800 who worked on it had been killed. Peter Colechurch's Old London Bridge was then demolished, and the remains of his body were found inside the chapel pier where Peter had been buried. Many Roman coins and wooden piles tipped with Roman type iron were dredged up at the time, which proved there had definitely been a bridge there in Roman times, although sited a little to the east. Now there is a new New London Bridge. It was opened by Elizabeth II on 16 March 1973 and has three spans of 260, 340 and 260 feet (79, 104 and 79 m.), and a width of 105 feet (32 m.). Rennie's New London Bridge has been rebuilt stone by numbered stone in Lake Havasu City, Arizona.

The ideas which produced the difference between Old (1209) and New (1831) London Bridges came originally from Italy. They spread to Paris

The modern London Bridge, opened in 1973 has three spans of 260, 340 and 260 feet (79, 104 and 79 m.).

where they reached their zenith in Jean Perronet's bridges. The Renaissance brought a new aesthetic ideal; art, science and technology came together after being separate for a thousand years. Late medieval experiments with segmental (less than semi-circular) and elliptical arches came to fruition and produced better shapes, more economic in their use of material, which challenged the semi-circular and pointed arches used so far. The gothic pointed arch is nearest to the ideal shape (a parabola) for an arch carrying an equally distributed load, but whereas this is fine for the great medieval cathedrals with their soaring vaulted roofs—over 100 feet (30 m.) up in the twelfth-century church of St Denis just outside Paris, and 158 feet (48 m.) in Beauvais Cathedral—a bridge does not want to be too high compared to its span because it makes the road inconvenient. To avoid this hump-back effect, very massive spandrels would have to be built. Thus the segmental and elliptical arches are better, not from a stress point of view in the voussoirs but because they require the least amount of material overall.

A bridge which might be regarded as a transition is the Ponte della Trinità in Florence. It was built by Bartolomeo Ammanati in 1569. There were three very flat pointed arches; their rise to span ratio is one to seven instead of the more usual one to four of that time. The foundations were very well constructed as Ammanati excavated 13 feet (4 m.) of the river bed before driving piles for a further 14 feet (4 m.). Large foundation stones were laid directly on the heads of the piles and the masonry piers built up from there. This work was done inside coffer dams. The spans were 87, 96 and 86 feet (26, 29 and 26 m.). The bridge which can be seen in Florence today is a faithful reconstruction using the original stones, as in 1944 retreating German troops blew up all the old bridges in the town, except the Ponte Vecchio.

The man who brought the masonry arch to perfection was Jean Rodolphe Perronet (1708–94). He was one of the first civil engineers, as distinct from an architect or a craftsman, and become the first director of the Ecole des Ponts et Chaussées in 1747 which was the natural extension of the Corps des Ponts et Chaussées constituted in 1716 by Louis XV to approve all new bridges and roads in France. Technically Perronet's best bridge was the Pont St Maxence not far from Paris where the Flanders Road crosses the River Oise. Its three arches were the flatest yet built: 6 feet 5 inches (1·95 m.) rise to a span of 72 feet (21·9 m.), that is, one to eleven. The very narrow piers were two columns joined by a lateral arch and each only stopped 9 feet (2·7 m.) of the stream. The bridge stood for nearly a hundred years until the Germans destroyed it in 1870. Perronet used his piers to take only the dead weight of the arches, and he let the side thrust of one arch at a pier be balanced by the side thrust of the other arch, which had to be the same span. In this way all the side thrusts were transferred to the abutments at each end of the bridge, and the piers could be much thinner offering less resistance to the river. Of course, if one arch were broken all the rest would fall down because they were leaning on one another, and for the same reason all the wooden centrings had to be left in position until all the arches had been set.

The Pont de la Concorde in Paris was Perronet's last bridge. He was seventy-eight then and this bridge was to be his best. He designed very flat segmental arches on slender doric columns for piers. He moved into a pavilion on the site and all was ready to start when ignorant or jealous people in authority ordered that the piers should be solid and thicker, and that the arches should not be so flat, but it is still a superb bridge. The five spans of 102 feet (31 m.) each have a rise of 13 feet (3·96 m.). Perronet was still in the pavilion at the end of the bridge when he died, three years after the bridge was finished in 1791.

Ironically Perronet achieved all this at a time when masonry bridges were just about finished with. A new age was coming and iron was the new material. Indeed the first cast iron arch in the world had already been erected at Coalbrookdale, Shropshire in 1779. Abraham Darby and John Wilkinson, ironmasters, decided to build an iron bridge over the River Severn to replace the ferry. Wilkinson was said to be "iron mad" because he

Coalbrookdale Iron Bridge heralded a new era in 1779.

believed that ships and houses could be built of the metal. The ironmasters employed an architect to make a design but rejected it because he proposed to use masonry, with iron playing a subsidiary role. It was Darby who made the final design. The bridge is a single 100-foot (30-cm.) span, semi-circular arch of five cast iron ribs supporting a 24-foot (7·3-m.) wide roadway made of flat cast iron plates. The total weight is 387 tons (393,200 kg.). All the castings were made in open sand moulds and each rib was cast in two halves. The prefabricated parts were taken in barges to the site and erected in three months with no trouble. Originally the abutments were solid earth-filled masonry, but this proved too heavy and timber side bridging was used until the 1820's when the present iron side arches were put in.

The next significant iron bridge was built by Thomas Telford 3 miles (5 km.) upstream at Buildwas between 1795 and 1798. It was a single 130-foot (39·6-m.) segmental arch with a rise of 27 feet (8·2 m.), and although it was longer than the Coalbrookdale iron bridge it used less than half the iron.

The masonry arch had served mankind well for nearly 2,000 years, but in the nineteenth century it had to give way to iron. These new beginnings by Darby and Telford were to lead up to the Sydney Harbour Bridge of 1932 which has a steel arch with a span of over 1,600 feet (500 m.).

5

Road Works before the Motor Car

Prior to the nineteenth century the roads in Britain were in a very poor condition. They were particularly bad in clay areas; carriages would sink to their axles and horses to their bellies. Daniel Defoe in 1722 observed a team of twenty-two oxen taking two years to carry a tree trunk from the forests of Kent to Chatham for boat building. Coaches bound for Holyhead were often dismantled at Conway in North Wales and carried piece by piece over the 15 miles (24 km.) to the crossing of the Menai Straits. There had been no workable system of road maintenance since the Romans left. The communities were small, self-sufficient and had little need of roads. The Church was the only central authority left and it did carry out some maintenance work, but not enough to keep up with the rate of decay. In Tudor times the secular power of the King was increasing and he felt the lack of communications. The first Highways Act was passed in 1555 which appointed an honorary surveyor for each parish, who in theory could demand four days' work from each person, but he had little support, money or knowledge, so that the potholes of the previous autumn and winter might or might not be filled up in the springtime. Then Acts were passed which in effect tried to ban traffic, restricting wheel sizes, numbers of draft animals, etc. The toll system was tried next and provided some answers to the problem.

The first toll gates were erected in 1663 on the Great North Road (London to Edinburgh), but the main turnpike (a bar of wood with spikes designed to stop cavalry) system, in which a section of road was turned over to a trust, did not come in until 1706, again on the Great North Road at Wadesmill in Hertfordshire. A turnpike trust was a body of men who undertook to build and maintain a road in return for tolls, usually on a non-profit-making basis. By 1750 there were 169 Turnpike Trusts, 530 in 1770, and in 1830 1,100 responsible for 23,000 miles (37,000 km.) of road. Unfortunately, most were badly managed, either through fraud or incompetence. Road making or mending often consisted of digging loose earth from the ditches and piling

A list of tolls on a toll-house in Angelsey, North Wales on the Holyhead Road.

it on the roadway in the hope that the traffic would beat it flat. If rain came before the traffic the road turned to mud. Gravel was sometimes applied and it gave a surface for a while but eventually it mixed with the mud. However, some progress was made. Now at least some main roads were maintained by professionals like Metcalfe, Telford and Macadam, instead of by amateurs as under the statute labour system. In 1734 it took at least nine days to get from London to Edinburgh; in 1830 the journey required only forty hours.

A lot of Britain's goods could travel by navigable rivers and in coastal waters. France, having less coast line, had preserved its Roman roads better. Even so, the first new roads since the Romans left were not built until Napoleon wanted to dominate Italy, and he constructed his roads through the Alps. The road over the Simplon Pass was a little-used mule track when in 1801 Napoleon brought in Nicolas Céard as his engineer to survey it for a road 16–20 feet (5–6 m.) wide. Six thousand men went to work on the 38-mile (61-km.) road in the next four years. It is estimated that accidents caused up to ten per cent of them to die; certainly over a hundred were killed in the Gondo Ravine working day and night through solid rock. It was traversed in part by a tunnel 731 feet (223 m.) long where all the holes for the blasting charges had to be hand drilled with hammer and chisel. The steepest gradient was 9:100. At the same time Napoleon's engineers were putting roads through the Mont Cenis and Mont Genèvre Passes. The Mont Cenis Road, finished in 1810 but usable in 1805, was the widest trans-alpine road until the 1930's.

Napoleon could build the best roads in the world because they were necessary for his military campaigns, and he had a centrally organized pool of men and scientific knowledge. This was the Corps and the Ecole des Ponts et Chaussées, formed in 1716 and 1747 respectively to oversee and study the building of bridges and roads. In 1764 Pierre Trésaguet, one of their engineers, laid the first proper road metal since Roman times. It had a foundation of large stones on edge, parallel in profile to the final crown of the road. These were covered by a 7-inch (18-cm.) layer of coarse broken stones, set by hand, and a top layer of smaller stones about the size of walnuts. After traffic use, the stones consolidated into a waterproof metalled road which shed its rainwater into side ditches.

"Blind" Jack Metcalfe was the first to apply Trésaguet's methods in England. A turnpike road was to be built from Boroughbridge to Harrogate, Yorkshire, and Metcalfe offered his services in 1765. He was a carrier and knew the topography of the land intimately, despite being blind after suffering an attack of smallpox at the age of six. He went on to build 180 miles (290 km.) of road in that area. Thomas Telford's roads were metalled in essentially the same way, except that he started from a flat base using more stone in the middle of the intermediate layer to give the surface a crown of 4 inches (10 cm.). The intermediate layer was over 7 inches (18 cm.) thick of $2\frac{1}{2}$-inch (6·4-cm.) stones, packed by hand. He put a 2-inch (5-cm.) layer of clean gravel on the top. Trésaguet, Metcalfe and Telford built good and lasting roads but they were expensive to construct. Therefore the vast majority of nineteenth-century roads were macadamized.

John Loudon Macadam (1756–1836) was a moderately rich man whose hobby had been road making during his fifteen years as Deputy Lieutenant for the County of Ayr. He became a trustee for the Bristol Turnpike and then their surveyor in 1816 in charge of 149 miles (240 km.) of road. Within eighteen months they were in good condition and the Bristol Turnpike was paying off its debts. He gave advice freely to other turnpikes and at his own expense directed and trained their men. After 1823 Parliament voted him £10,000 (£8,000 of this was for past expenses) and he was appointed Surveyor-General of Metropolitan Turnpike Roads which led to the general adoption of his system of road building throughout Britain, and thence the world. Macadam believed that a road surface should be slightly elastic and

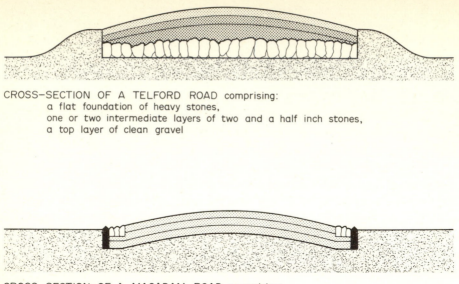

CROSS–SECTION OF A TELFORD ROAD comprising:
    a flat foundation of heavy stones,
    one or two intermediate layers of two and a half inch stones,
    a top layer of clean gravel

CROSS–SECTION OF A MACADAM ROAD comprising:
    two or three thin layers of one inch stones

Cross-section of various road surfaces.

that a well-drained sub-soil, under a rainproof surface, would act as a foundation. Thus his roads consisted only of 6–12 inches (15–30 cm.) of 1-inch (2·5-cm.) stones—"Any more would be mischievous," he said. Sometimes fine gravel was put on the top and the excess brushed off. Under the action of traffic all these various kinds of stone road consolidated into a waterproof slab. The reason was that the hard rimmed wheels broke off bits of stone and dust which were carried by rainwater into the interstices of the larger stones. The surface tension of the water bound the whole lot together; thus they had to be watered during long periods of drought.

However, there is more to a road than its surface. Its alignment has to give the shortest distance for the easiest gradients. This means that a careful survey must be made, and cuttings, embankments and bridges built. Thomas Telford (1757–1834) built some beautiful roads. He was trained as a stonemason, became an architect and worked as such in Edinburgh, London and Portsmouth, before his appointment as County Surveyor for Shropshire in 1787. In the three years following the storms and floods of 1795, he had to rebuild three bridges over the River Severn. The one at Buildwas was his first iron bridge, with a single span of 130 feet (39·6 m.). Not all his work was so spectacular. The churchwardens of St Chad's in Shrewsbury asked him to report on their leaking roof. After an examination of the building, he pointed out that the church tower was about to fall over due to the shallow medieval foundations. They curtly dismissed him and his report. Three days later the clock in the tower struck and the tower collapsed.

In 1801 Telford was commissioned to make a report on the dying High-lands of Scotland. "The Lairds have transferred their affection from the people to flocks of sheep," he said, because sheep farming was more profitable. Thus there was wholesale emigration. To alleviate this problem Telford recommended a programme of road and harbour building. In the next eighteen years he made 920 miles (1,480 km.) of new road, rebuilt 280 miles (450 km.) of old road and constructed 1,117 bridges. In some places there was no suitable stone for the bridges and it had to be brought for many miles by pack animal. Some bridges merely carried a small stream under his road while others were pieces of engineering in their own right. The longest is over the River Tay at Dunkeld having seven stone arches varying from 20 feet (6 m.) at the ends to a 90-foot (27-m.) span in the middle. To lessen the dead weight of his stone bridges, Telford built the spandrels, or sides, hollow with internal cross walls instead of the usual rubble filling which he described as a heap of rubbish. He used a very delicate tracery of cast iron for the 150-foot (46-m.) arches at Craigellachie over the River Spey. John Mitchell, Telford's deputy in Scotland, estimated that he travelled about 10,000 miles (16,000 km.) a year in connection with these works, most of it in very wild country. Telford himself made an extended tour once a year.

Telford's general specification for a main road was that it should be 34 feet (10 m.) wide between the fences with the central 18 feet (5 m.) metalled, as described earlier, and the remaining sides covered with gravel. At least every 100 yards (91 m.) there should be a drain running from the foundation to the outside ditch. The embankments and cuttings should be 30 feet (9 m.) wide at the road level with sides sloping at no more than two in three. The gradient of the road should not exceed one in thirty.

In the early nineteenth century several Irish M.P.s were pressing for better communications between London and Dublin. To this end the Port of Holyhead on the island of Anglesey had been built but nothing had been done to the Holyhead Road which was in such a state that the London mail coaches did not operate west of Shrewsbury. In 1815 Telford was asked to build or rebuild this road. The London to Shrewsbury section, 153 miles (246 km.), was improved, having its bends eased by short pieces of new route, its gradients lessened by making cuttings and embankments, and all its surface metalled. It was in the charge of seventeen turnpike trusts and they were allowed to stay provided they maintained the road to the new standard. But on the 107 miles (172 km.) from Shrewsbury to Holyhead the seven turnpikes were bought out by the government. West of Llangollen in North Wales deep cuttings had to be blasted through rock and large embankments constructed to maintain a gradient no steeper than one in twenty. The

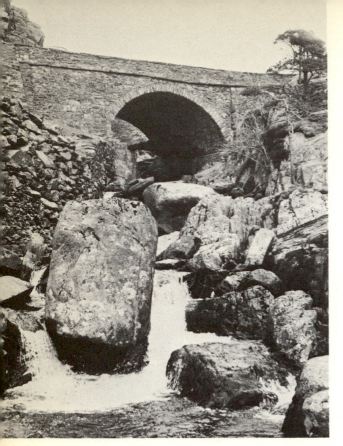

One of Telford's stone arches carrying his Holyhead Road over the River Ogwen in North Wales.

An old toll cottage on Holy Island at the end of the embankment that Telford built for his Holyhead Road.

summit of the road is in the Nant Ffrancon Pass. At the head of the pass the road is in a rock cutting, then it runs on an embankment leading over a stone arch under which the River Ogwen falls spectacularly to the valley below. The biggest embankment on the road joins Anglesey to Holy Island. It was made in 1823, 1,300 yards (1,189 m.) long, 16 feet (4·8 m.) high and 114 feet (34·7 m.) wide at its base. It has since been widened to carry the railway as well. The greatest single engineering work on the Holyhead Road, and the longest span bridge in the world at the time, was the suspension bridge over the Menai Straits between Angelsey and mainland Wales. The roadway is 579 feet (176·5 m.) long and 100 feet (30 m.) above the water.

The idea of a suspension bridge is very old. Such bridges were used by all the ancient societies, in China, India and South America, but the roadway usually followed the curve of the suspension cables, which were made of vines, bamboo rope or strips of hide. The first modern suspension bridges with a level floor and capable of carrying vehicular traffic were built by James Finley in North America. Of the forty-odd bridges made to his design, the longest was a footbridge built in 1809, 308 feet (94 m.) across the Schuylkill River in Philadelphia using chains of wrought iron links. Telford, almost certainly aware of Finley's work, had to use a suspension bridge for his Holyhead Road to cross the Menai Straits because the width of the channel and the clearance required for shipping precluded the use of an arch or arches. Telford decided to use sixteen chains made of 9-foot (2·7-m.) long, wrought iron eyebars held together by 3-inch (7·6-cm.) diameter pins, whereas Finley used wrought iron links. Each chain was five eyebars wide. The eyebars were tested to 10 tons (10,000 kg.), twice the load they would take on the bridge, on a special machine designed by Telford. Then they were heated and dipped in linseed oil to preserve them.

The site chosen for the bridge was known as Pig Island and at the eastern end of the Straits. Pig Island was a rock just off the Angelsey shore from which

An illustration of the Menai Straits Suspension Bridge taken from Telford's autobiography. The bridge was finished in 1826 and is still in use, except that the chains and road deck have been renewed.

drovers used to swim their animals across at low tide. Work started in 1819 by blasting this rock flat to act as a base for the Angelsey tower. The tower on the mainland side was founded on rock 6 feet (1·8 m.) below the low water line. William Provis was the Resident Engineer with 200 men in his charge. By 1821 the towers were 60 feet (18 m.) up and the anchorages for the chains finished. The Angelsey anchorages were made by driving two 6-foot (1·8-m.) diameter tunnels into the rock for 20 yards (18 m.) and then joining them by a cross tunnel, in which a cast iron frame was placed to hold the ends of the chains. On the mainland side there was no rock near the surface so the chains went much further back to find the rock through tunnels filled with masonry. The bridge is not symmetrical for this reason. In 1825 all the masonry was finished: the towers up to their full height of 153 feet (46·7 m.), and the approach viaducts, three arches on the mainland side and four on the Angelsey side, each of 52-foot (15·8-m.) span. The side chains were in position and part of the centre span chains hung down the Welsh tower to the water level.

On 25 April Telford decided that the weather would allow him to lift the first of the centre chains. This was carried on a 450-foot (137-m.) long raft and pulled by four boats into position across the channel at high slack water. One end of the eyebar chain was fixed to that hanging down the Welsh tower and the other to 6-inch (15-cm.) diameter ropes going over the Anglesey tower to capstans worked by 150 men. They pulled the first centre span chain up in one and a half hours. There were cheers when Telford and

Provis standing on the tower waved their hats to signal that the connection had been made, and in the jubilation that followed three men walked across the 9-inch (23-cm.) wide chain. The other fifteen chains followed when conditions were right. The last one went up on 9 July 1825, when a band played the National Anthem on a wooden platform attached to the chains 100 feet (30 m.) over the centre of the Straits. The chains were arranged in four groups of four. All that remained now was to build the wooden roadway. This was 30 feet (9 m.) wide, 579 feet (176·5 m.) long and held by suspenders every 5 feet (1·5 m.).

The Menai Bridge was opened in January 1826. The timber deck was replaced in 1893 by one of steel designed by Benjamin Baker. The wrought iron eyebars of the chain remained in use until 1939, when they too were replaced by steel.

The roads in the United States in the nineteenth century were much worse than in Britain. In fact, apart from town streets, there were only two roads of any length that were properly metalled. These were the Lancaster Turnpike and the Cumberland, or National, Road. Even in 1915, eighty-seven per cent of the roads were native earth.

In 1818 the Cumberland Road ran from Cumberland on the Potomac River in Maryland over the mountains to the Ohio Valley where it terminated at Wheeling. The 112 miles (180 km.) had cost twice the estimate because of incompetence and fraud and other extras not anticipated. Army engineers rebuilding the road in 1834 pointed out where the specifications had not been met. It was planned to be 66 feet (20 m.) wide with the middle 20 feet (6 m.) metalled, there were to be ditches on both sides and no gradient steeper than nine in a hundred. It was sometimes called the National Road because an Act of Congress in 1806 said that it should be built at public expense, and as the only metalled road to the west it was used a great deal. After 1850 it crossed the Ohio River, when first Ellet and then Roebling built the suspension bridges at Wheeling (mentioned in Chapter 12), and it went on for a total of 834 miles (1,342 km.) through Ohio, Indiana and on into Illinois. A Finley-designed suspension bridge carried the Cumberland Road for a while, over Dunlap Creek, Uniontown, Pennsylvania. It collapsed in 1820 owing to the weight of snow lying on it, and it was replaced in 1839 by the first iron bridge in the United States. This was designed by Richard Delafield, and G. W. Cass was in charge of its erection. The bridge is still in use and has an arch span of 80 feet (24 m.). There are five ribs each made of nine cast iron box voussoirs bolted together. The road is 25 feet (7·6 m.) wide and about 40 feet (12 m.) above the water.

In 1795 the Lancaster Turnpike ran 62 miles (100 km.) from Philadelphia to Lancaster through the farming country of Pennsylvania. As its name suggests, it was a toll road and it yielded six to fifteen per cent dividend until the railroad came. David Rittenhouse, a scientist, did the original surveys but the road was hastily and badly made and it soon broke up. William Weston, an English engineer building canals in the area, helped to make the road good. It was 20 feet (6 m.) wide and 17 inches (43 cm.) deep made of stones overlaid with gravel, but this was not a macadam surface. The first road in the United States to be macadamized was the Boonesborough Turnpike in 1823, running 8 miles (13 km.) to Hagerstown, Maryland. The Lancaster Turnpike crossed the Schuylkill River in Philadelphia by a bridge of planks resting on floating logs. Timothy Palmer replaced this in 1805 with his Permanent Bridge, made of wooden arch trusses, which carried the Lancaster Turnpike for fifty years. It had a central span of 195 feet (59 m.) and two side spans of 150 feet (46 m.). The western pier went down 42 feet (13 m.) to bedrock inside a coffer dam designed by William Weston.

The wooden truss was very popular in the pioneering days of the United States because the materials were easily obtainable and such bridges could be made by any competent carpenter. The continuation of the Lancaster Turnpike had to cross the main channel of the Susquehanna at a place known as McCall's Ferry. In 1815 Theodore Burr erected a single 360-foot (110-m.) timber arch over the 100-foot (30-m.) deep swift-flowing river. He built the truss in sections a quarter of a mile (400 m.) downstream having considerable trouble with storms that nearly carried them away. He waited for the river to freeze and erected his arch from scaffolding on the 60–80-feet (18–24-m.) deep slowly moving mush of broken ice. The ice had been broken by rocks upstream and forced into the narrows at McCall's Ferry. The arch was successfully set up with only one man being hurt; he fell 54 feet (16 m.), struck the braces twice and tumbled into the icy water. He was at work again in a few days. The ice destroyed the bridge after two years and it was not replaced.

From these beginnings American engineers developed the modern truss girder. Ithiel Town patented a diamond pattern truss in 1820. It was straight with vertical ends and he made a fortune from the $1 or $2 per linear foot he charged for using his idea. The longest Town truss was 2,900 feet (884 m.) on eighteen piers over the River James at Richmond, Virginia. Others followed. William Howe in 1840 used wrought iron for the tension members and in 1844 Caleb and Thomas Pratt improved upon this and produced the type of iron truss girder still in use today. A very important contribution was made by Squire Whipple in an *Essay on Bridge Building* in 1847. He appears to have been the first engineer to consistently apply

66

A packhorse train. This method of transport was by far the most reliable on pre-Macadam roads.

mathematics to find the stresses in the lattice work of struts (in compression) and ties (in tension) in a truss girder. It was now possible for an engineer to calculate beforehand the stresses in the various members of his bridge without having to rely solely on his intuition.

The works described in this chapter certainly improved the roads of the eighteenth and nineteenth centuries beyond all recognition, but road transport itself could not become a viable proposition until the internal-combustion engine had been invented and motor vehicles developed in the twentieth century. Once the industrial revolution had got under way, horses and carts and pack animals were just too slow, and the loads to be shifted too many and too heavy. Canals, and then railways, were the significant transport systems of the eighteenth and nineteenth centuries.

6

Canals: Three Pioneers

The three most famous canal engineers are James Brindley (1716–72), Thomas Telford (page 60) and Ferdinand Marie de Lesseps (1805–94). The first has been overrated and the last was not an engineer at all but a diplomat. However, Telford was probably the greatest civil engineer ever.

The misconceptions about Brindley arise from Samuel Smiles's book *Lives of the Engineers* published in 1862. Smiles was obsessed with the idea called "self-help": that success always comes to those who work hard. Thus it suited his purpose to have Brindley start life as an illiterate peasant and finish as a sophisticated professional engineer, having been given his opportunity by Francis Egerton, third Duke of Bridgewater. In fact, Brindley had an extensive engineering practice before he met the Duke and was already surveying the Trent and Mersey Canal in 1758. The Bridgewater Canal, however, made him into a civil engineer of national repute, but he remained essentially an intelligent artisan. His canals were no advance on the French canals of the seventeenth century mentioned in Chapter 3, but they had a much greater effect because Britain was just starting her industrial revolution whereas France was not.

The Duke of Bridgewater owned coal mines at Worsley in south Lancashire and to avoid the inadequacies of land transport he and his agent John Gilbert decided to build a canal to Salford. James Brindley was to build it for them. Which of these three men was really the guiding light for the design is not known, but Brindley's subsequent canals are nowhere near as daring in their engineering as the Bridgewater. The original part of the canal started inside the mines at Worsley. After 10 miles (16 km.) without locks, it ended at Castefield, Salford in a tunnel. The power of the canal water falling into the River Medlock was used to work the hoist lifting the coal in containers from the barge, up a shaft and directly into the market. The canal crossed 40 feet (12 m.) above the River Irwell on the Barton Aqueduct. John Smeaton (page 42) thought this to be impracticable, but Brindley built it with a water channel 18 feet (5·5 m.) wide and $4\frac{1}{2}$ feet (1·4 m.) deep.

He had some trouble with the approach embankments in the bogs of the Irwell Valley, which he solved by digging ditches to drain and dry the foundations. The price of the Duke's coal was halved in 1761 when the canal opened, and he made a fortune. Brindley made his name and was then in great demand as a canal engineer.

The mine entrance at Worsley, the start of the original Bridge-water Canal in 1761.

Barton Aqueduct which carried Brindley's Bridgewater Canal 40 feet (12 m.) above the River Irwell.

Brindley's biggest single work, and the most extensive was the Trent and Mersey Canal. Work started on it after Josiah Wedgwood, the chief promoter, cut the first turf and roasted a sheep in Burslem for the poorer potters on 26 July 1766. In its 93 miles (150 km.) it has 76 locks, 213 over-bridges, 160 aqueducts and 5 tunnels. Brindley lined all his canals where necessary with puddled clay, that is, clay and sand kneaded together, trodden flat and dried in layers. Most of the aqueducts are small but that over the River Dove has twelve 15-foot (4·6-m.) spans to make an overall length of 267 feet (81·4 m.) clearing the river by 6 feet (2 m.). All Brindley's aqueducts are heavy and moderately low because of the puddled lining. He had the River Dove dug out to twice its width and diverted it to one side while he built six arches of the aqueduct in the dry river bed. The other half was built with the river flowing through the completed half.

The longest of the tunnels on the Trent and Mersey Canal is at Harecastle in Cheshire, $1\frac{3}{4}$ miles (2·8 km.) long and 9 feet (2·7 m.) wide. Much longer tunnels had been built for centuries in the mines, but Brindley must be regarded as a pioneer tunneller because transport tunnels are different

Harecastle Tunnels, Brindley's on the right and Telford's on the left.

from mine tunnels. Mine tunnels follow a seam and generally bear no significant relation to the surface, but a transport tunnel has to follow a direct line. Harecastle Hill was surveyed and the line of the tunnel staked out. Fifteen shafts were put down up to 190 feet (58 m.) to the tunnel level so that, working from the bottom of each shaft and each portal, the tunnel was attacked on thirty-two faces simultaneously. Even so Brindley met many obstacles and the tunnel took eleven years to complete. Pumps to remove the water encountered were driven by water or animal power, but as the tunnel progressed more and more water came in, until Brindley had to use steam pumps. The rock was very hard and had to be blasted. Dangerous gases escaped from coal seams crossed by the tunnel which meant that a ventilation system had to be devised; Brindley put a brazier of red-hot coals at the bottom of a stove pipe in the shafts to set up convection currents. He had bought a quarter-share of these coal seams in 1760 very cheaply, and after the tunnel was completed, the mine was opened using branch tunnels into the main tunnel for the transportation of the coal. With the Harecastle tunnel finished in 1777, the Trent and Mersey Canal was operative over its full length.

James Brindley died of diabetes in 1772 and his many assistants, Robert Whitworth and Hugh Henshall in particular, finished his works. He was responsible for over 500 miles (800 km.) of waterway, 298 locks, 847 bridges and 12 tunnels. He said that the purpose of rivers was to supply water for canals. He represents the birth of civil engineering as a profession, because he started to separate the roles of engineer and builder (contractor). There were no contractors then as we know them; the contractor in the eighteenth century was more like a foreman of a gang who would do a job for piecework payment. After Brindley's death, Thomas Telford went on to establish the profession and the system of Chief Engineer, Resident Engineer, and Contractor. The contractors had to await the railway era before assuming their present importance.

By 1820 Harecastle Tunnel was a serious bottleneck and it was worn out. In places its roof sagged by 2 feet (60 cm.) because of mining subsidence, and elsewhere its lining was rubbed away. Telford was called in to build the inevitable second and parallel tunnel 25 yards (23 m.) away from the first. The contractor was Daniel Pritchard who started work in the summer of 1824 by sinking fifteen shafts and making sixteen cross-headings into the old tunnel. They met quicksand near the north end which had to be pumped out, then reached the same very hard rock that Brindley had encountered which had to be chopped and blasted. The new tunnel is wider and higher and a little longer, but it was finished in three years, a good measure of the progress of civil engineering in the sixty years that had elapsed since the building of the old tunnel.

Thomas Telford first proved himself to be a daring engineer capable of original solutions to new problems with his Pont Cysyllte Aqueduct on the Ellesmere Canal, "the greatest work of art" Sir Walter Scott had ever seen. The Ellesmere Canal was to link the present Ellesmere Port on the Mersey with Shrewsbury on the River Severn. It never achieved this but the crossing of the Vale of Llangollen at Vron Cysyllte was planned with this in mind. Telford was appointed to the Company in 1793 under William Jessop who had carried out the survey. Jessop, a then famous canal engineer, taught Telford all he could about waterways but he soon recognized a superior constructional engineer. The Pont Cysyllte is a 1,007-foot (307-m.) long trough made of cast iron plates. It is held 127 feet (39 m.) above the River Dee on eighteen masonry piers, giving nineteen spans of 53 feet (16 m.) each. The channel is 11 feet (3·4-m.) wide with a towpath over one side. The whole thing is breathtaking, viewed from above or below. The southern approach embankment is 97 feet (30 m.) high and was the biggest earthwork at that time.

The site was cleared in January 1795, then the canal committee wondered if it was all possible. Jessop and Telford assured them that it was, so in July the foundation stone was laid, in the first pier south of the river. Matthew Davidson was the resident engineer and he was to serve Telford again during the construction of the Caledonian Canal across Scotland. From 70 feet (21 m.) up, the piers were hollow with internal cross walls. They had to be built by highly skilled masons, John Wilson and John Simpson, who were also to feature in some of Telford's subsequent works, and were finished by 1802. The iron work, made at nearby Plas Kynaston Ironworks, could then be installed. All was not finished until the summer of 1805 because there had been financial delays.

Large-scale civil engineering is expensive, and to avoid this Brindley's canals followed the contours of the land and as a result were very indirect. A good example of this can be seen in the Oxford Canal where it crosses the River Swift. The valley is only a quarter of a mile (400 m.) wide, but the canal goes 1 mile (1·6 km.) upstream to cross on a low short aqueduct before it resumes its original direction. This was not a disadvantage for the water-powered industry of the time, but it was to the later steam-powered industry which was concentrated in the towns. The second generation of canal engineers built as straight as possible with massive embankments and cuttings. Telford's last canal, the Birmingham and Liverpool Junction, was built on this principle.

The Pont Cysyllte Aqueduct takes the Ellesmere Canal across the valley of the River Dee. It is 127 feet (39 m.) high.

The cast iron trough of the Pont Cysyllte Aqueduct is 1,007 feet (307 m.) long and 11 feet (3·35 m.) wide.

The first part of the new route was along Brindley's Birmingham Canal to Wolverhampton, which Telford saw as "a crooked ditch" before he rebuilt it, knocking 8 miles (13 km.) off its length. He made the summit level at Smethwick into a 70-foot (21-m.) deep cutting over which he put the lofty iron span of Galton Bridge. After Wolverhampton 39 miles (63 km.) of new canal were cut straight to the Chester Canal at Nantwich, containing every facet of civil engineering, including trouble. A 690-yard (630-m.) tunnel was planned at Cowley near Gnosall and construction began in the summer of 1830. Ninety yards (82 m.) from the north end a fault was struck. They went on for another 150 yards (137 m.) through very frail rock and then had to stop. Telford decided to open out the tunnel into a cutting where the rock was bad and finished up with a tunnel 81 yards (74 m.) long. There were notable cuttings at Tyrley and Grub Street. Tyrley cutting, 90 feet (27 m.) deep and $1\frac{1}{2}$ miles (2·4 km) long, was built in weak rock alternating with clay. There were many rock falls and slips, and the sides had to be cut back and retaining walls built in places. Grub Street near High Offley was the same but 2 miles (3·2 km.) long.

There were many embankments but the one which gave the most trouble need not have been made. It came about because Lord Anson had a game reserve in Shelmore Wood on the direct line of the canal and at its level. The Lord would not sell so Telford had to design a by-pass for the canal on a 1-mile (1·6-km.) long embankment 60 feet (18-m.) high. Four hundred men and seventy horses moved in during 1829 and started to tip spoil from Grub Street cutting and Cowley "tunnel" to form the great bank. In a year, half a million cubic yards had been dumped and the embankment was up to 45 feet (14 m.), but it was slowly sinking and spreading along its whole length. They carried on hoping to reach an equilibrium. In August 1832 there was a big slip 800 yards (732 m.) long and half the width of the bank, and it was the only part of the canal unfinished. Telford was seventy-five and ill so William Cubitt was appointed as his deputy. Cubitt was soon to be famous as a railway engineer, but he cut his teeth on Shelmore Bank. For two years he dumped sand and hard core to dry out and consolidate the bank. "It takes more earth than even I had anticipated," he reported to the Canal Company. In March 1834 Telford viewed the bank and Lord Anson's pheasants snug in Shelmore Wood for the last time. In May there was another slip, and 10,000 cubic yards of rock fell into the Grub Street Cutting, obliterating all signs of the canal for 60 yards (55 m.). In September Telford died. The Birmingham and Liverpool Junction Canal was eventually opened for traffic along its whole length in January 1835.

Telford lived to a remarkable age (seventy-seven years) for an engineer of those times, as not many saw sixty years. He was the first president of the Institution of Civil Engineers in 1820. He set the high standards for the profession which he forged, bringing bold originality, and social responsibility.

In judging the merit of an engineering feat, the economic effects of the work and the daring of its perpetrators must be considered as well as the technology involved. The Suez Canal is worthy of consideration because it knocked 4,000 miles (6,400 km.) off the 10,000 miles (16,000 km.) by sea from Liverpool to Bombay, and Ferdinand de Lesseps overcame or ignored immense political pressures against its construction from Britain and Russia. From an engineering point of view the canal could have been built by anyone in the last five thousand years, although it was the first of the "ship" canals. It is only 40 miles (64 km.) of ditch linking 60 miles (96 km.) of natural depressions, now lakes, and the highest land was only 36 feet (11 m.) above sea level and mostly sand.

Indeed, a water link between the Mediterranean and Red Seas had been made several times before. One of the branches of the Nile Delta used to enter the Mediterranean near to the modern Port Said and to link this with the Red Sea was no trouble to the ancient Egyptians. The Pharaoh's route was eastwards from this old branch along a linear depression in the desert to join Lake Timsah, then south through the Bitter Lakes to the Red Sea. After the Romans occupied Egypt, Trajan reopened the canal in A.D. 98 and it was in use again by Arab rulers around A.D. 700. These early canals were probably only navigable at the time of high Nile and were not important through routes. Traces of them were found by Napoleon's engineers and by the builders of the modern canal.

Napoleon sent his surveyor Lepère to the Isthmus who reported that the Red Sea was 32 feet (9·7 m.) higher than the Mediterranean. Actually there is virtually no difference in the levels, as a team of British engineers reported in 1833, but Lepère's error killed the idea for Napoleon. In 1833 a group of French philosophers led by Prosper Enfantin commissioned a survey. The job of chaperoning the party was given to the French Vice-Consul in Alexandria, Ferdinand de Lesseps. He became interested in the canal but had to wait until 1854 and the succession of Mohammed Said as Viceroy of Egypt, for the canal to be considered seriously. De Lesseps skilfully bullied, flattered and inflated Said; he diplomated his way round Europe pulling together the previously hesitant supporters of the project and charming or ignoring the political opposition, until on 25 April 1859 he ceremonially struck the first sod on the strip of land that separated the Mediterranean from Lake Manzala. The plan for construction was essentially that drawn

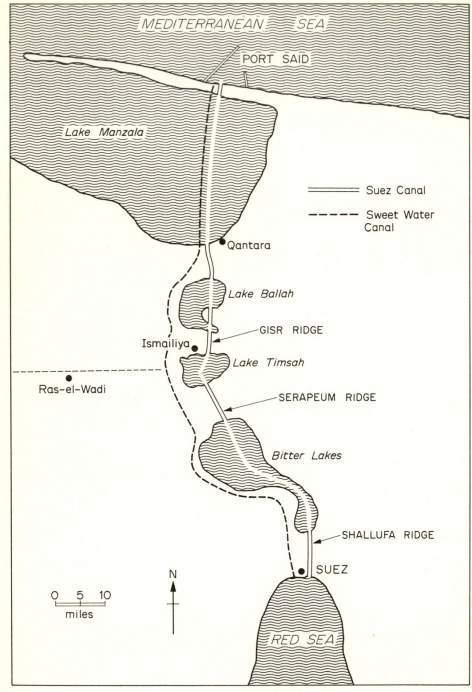

Map of the Suez Canal.

up thirteen years before by Linant de Bellefonds, a French engineer working for the Egyptian Government. It had three sections: the Sweet Water Canal, Port Said, and the Maritime Canal itself.

The Isthmus of Suez was pure desert and 80 miles (129 km.) east of the Nile; there was not a drop of water anywhere. The Sweet Water Canal was to provide this, but it was no mean water conduit. It was 60 feet (18 m.) wide and 8 feet (24 m.) deep. It was also a transport link for men and materials. The Egyptian Government built the 56 miles (90 km.) from the Nile to Ras-el-Wadi, and the Canal Company built the remaining 20 miles (32 km.) to Lake Timsah which was reached by February 1862. The canal then turned south and Suez received fresh water in December 1863. At this time Egypt was essentially a feudal country and 80,000 forced labourers working by hand were employed in the Sweet Water Canal's construction. Some had to be shown how to use wheelbarrows because never having seen them before, they carried them on their heads. The northern branch, not completed until 1869, was meanwhile supplied with water by pipeline. There were also distillation plants at Port Said to convert sea water to fresh water.

One of the partially-constructed breakwaters at Port Said.

The building of the artificial Port Said was the hardest task and not made easier by its inaccessible site. A narrow strip of land, 40 miles (64-km.) long, separated the Mediterranean from Lake Manzala and it was on this that Port Said was created by building two breakwaters and dredging this 450-acre (1,820 million sq. m.) harbour to 16 feet (5 m.). Over a mile (1·6 km.) in length, the breakwaters were built of stone quarried at Alexandria and from thirty thousand concrete blocks each weighing 22 tons (22,400 kg.) and made on the site. In all 142,000 square yards (118,000 sq. m.) of land were reclaimed.

The actual maritime canal was a relatively simple job. It involved cutting a channel through the ridges and flooding the depressions. Moving south from Port Said there were 26 miles (42 km.) through the shallow lagoon Lake Manzala to Qantara. The water here was only 5 feet (1·5 m.) deep so it was necessary to dig a channel and build retaining banks. A pilot channel 12 feet (3·66 m.) wide was made by serfs scooping out the mud in baskets, pressing it against their chests to squeeze out the water and laying it out in lumps. Then dredges were brought in and they reached down below the mud to stiff clay which, when sun-baked in layers, made strong banks 6 feet (1·8 m.) above the water level. The work was delayed and undone by occasional severe storms. The canal, 325 feet (99 m.) wide, was through the lake by 1866.

Behind Qantara there was an easy 3 miles (4·8 km.) of low-lying dry sand to cut through to reach Lake Ballah which was then a swampy depression. The spoil excavated in the lake was gypsum and unsuitable for the banks because it cracked, so material had to be brought in, presumably via the canal from Lake Manzala. The canal goes for 8 miles (13 km.) across Lake Ballah and was virtually finished this far by 1867.

Next, lying across the path of the canal were the sandhills of the Gisr Ridge, a major obstacle 9 miles (14 km.) long and rising to 36 feet (11 m.) above sea level. The method, as with all the dry land work, was to excavate a pilot channel by hand until there was enough water to float the dredges. The Gisr Ridge was not finished until 1866 but it was passable in 1862 so that the waters of the Mediterranean could flood the 4-mile (6·4-km.) long depression of Lake Timsah. On this occasion de Lesseps was enrolled on the French Legion of Honour.

Then came the Serapeum Ridge which ran for 10 miles (16 km.) and up to 29 feet (9 m.) above sea level. It had to be pierced before the canal could reach the Bitter Lakes. A little rock was encountered and it was not possible to get the manual labour to take the pilot cutting all the way to water level, so the cutting was dammed at each end and water admitted through side cuttings from the Sweet Water Canal, which is 17 feet (5 m.) above sea level. The dredges went from Timsah up locks into the Sweet Water and into the embryonic Serapeum Cutting. The Bitter Lakes were flooded in March, 1869 and formed a 25-mile (40-km.) stretch of water. This was probably long ago part of the Red Sea until the 13-mile (21-km.) Shallufa Ridge was formed north of Suez. Fifty-two thousand cubic metres of rock had to be blasted out of the Shallufa Ridge before the waters of the Mediterranean and Red Seas could mix in the Bitter Lakes on 15 August 1869. Port Suez was already a small port for the overland route to Cairo but had to be enlarged by building an 850-yard (777-m.) breakwater out to sea, dredging a channel and reclaiming land. The population of the town rose from four thousand to twenty-five thousand.

Early work on the Suez Canal showing the heavy reliance upon manual labour.

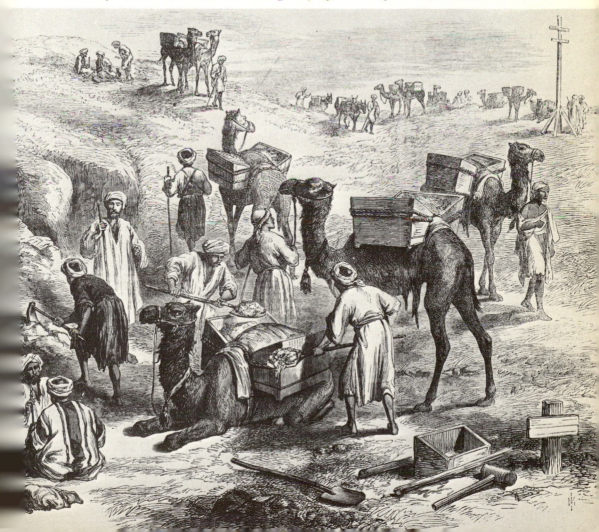

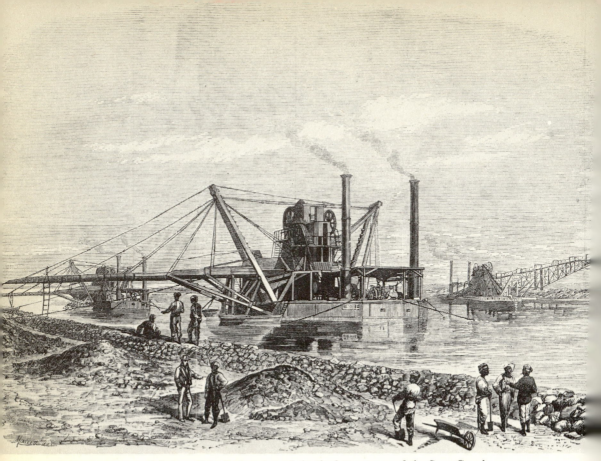

A long couloir dredge at work in the later stages of the Suez Canal.

Mohammed Said died in 1862 and the new Viceroy was Ismail. He was a progressive sophisticated young man and de Lesseps had to handle him differently from Said. Ismail was keen to bring Egypt into the nineteenth century, and his first act in power was to abandon the corvée or forced labour. This was a set-back for the Canal Company, who had been paying these twenty thousand and more labourers sixpence a day, as they now had to pay four times as much for unforced labour. The principal contractor, the French firm of Borel and Lavalley, had to design and import bigger and better dredges to replace the manual labour. These machines were in full operation by 1865; the most important were the "long couloir" and the "elevateur". Both were chain bucket dredges but the long couloir disposed of its spoil by sluicing it along a 100-foot (30-m.) chute 40 feet (12 m.) above the water, and the elevateur used a 200-foot (60-m.) double tramway carrying containers full and empty. The latter was used in cuttings where the banks were high.

The canal was virtually finished in time for the formal opening on 17 November, 1869. It was at its full depth of 26 feet (8 m.) except for a rock

ledge in the Shallufa Cutting. The bottom width was 72 feet (22 m.) minimum and the surface width varied from 196 to 327 feet (60–100 m.). The
task of freeing and modernizing Egypt was proving too much for Ismail.
Despite his early efforts the country was going towards bankruptcy through
mismanagement and corruption, and perhaps as a final bluff or just to join
in, Ismail went on a delightful spending spree. He toured Europe in 1869
inviting everyone of any consequence to the opening ceremonies of the canal
and the most extravagant party of the century. The principal guest was
Eugénie, Empress of the French. The Emperor of Austria, Francis Joseph,
came himself while Prussia was represented by the Crown Prince, and
Holland by the King's brother Prince Henry. Britain and Russia, dis-
approving, sent their ambassadors in Constantinople. A thousand people
were considered important enough to have all their travelling expenses
paid and tens of thousands came, invited or not. All the monuments in Cairo
were painted red and white irrespective of what delicate arabesques would
be covered. Eugénie arrived at Alexandria on 19 October and the great
celebrations started in Cairo. The Egyptian corvette *Latif* was sent down
the canal the day before the opening convoy of 17 November to make sure
the way was clear. It ran aground and Ismail gleefully suggested blowing it
up, but it was towed clear. The convoy took three days to travel through
the canal stopping a night at Ismailiya where a twenty-four-course dinner
was served. They arrived in Suez on 20 November 1869 and then went
back to Cairo for more entertainments.

Water transport is by far the cheapest in terms of tractive effort required
per unit of load carried, because a water surface is smooth and the friction
between it and a boat is small. It is therefore surprising to find that inland
canals are now largely neglected as a commercial transport system. If the
goods carried are non-perishable, then the slowness does not matter.
Generally, the only type of canal which still makes money is the larger ship
canal for ocean-going vessels. The first to be built was the Suez Canal and
it is not now in operation solely for political reasons, not commercial.

7

Railway Engineering: Three Pioneers

Railway engineering is embankments, cuttings, bridges and tunnels. These items had all been built before the coming of the railways but what makes railway engineering spectacular is that a lot of work was done in a short space of time to very great effect. Railways are essential for industrialization and easy cheap transport finally broke the old immobility of society. They also advanced the techniques of civil engineering and changed the structure of the industry in that the contractor gained in importance and the engineer receded. The construction of railways produced the first modern style contractor: Thomas Brassey.

Industrial revolutions, first in Britain and then elsewhere, demanded fast dependable transport, and could afford it. Three things are essential for great engineering: necessity, money and organization. All were present after the success of the Manchester–Liverpool Railway in 1830. They produced in England three engineers foremost among others.

Robert Stevenson (1803–59), with his father George, was the pioneer of the modern railway and therefore was the most influential of the three. He was careful and methodical in his engineering.

Isambard Brunel (1806–59) was an artist and a visionary. He did not accept traditional engineering principles without question. For example: his Great Western Railway was on a 7-foot (2·13-m.) gauge because he reckoned this was the most suitable for economic high speed running, whereas the Stevensons' standard 4-foot 8½-inch (1·43-m.) gauge was based originally on that used in some medieval German mines. But Brunel was ten years too late and the standard gauge prevailed. On the other hand, the size of his steamship *Great Eastern* was thirty years ahead of its time. His enthusiasm and confidence led him to make commercial mistakes, but with style.

Joseph Locke (1805–60) is less known. His philosophy of railway building was completely opposed to that of the Stevensons'. He believed that a railway should be built for the minimum viable cost and, therefore, he avoided heavy and costly civil engineering works if possible, and he did not mind the much steeper gradients that resulted. For example: his line between Lancaster and Carlisle in 1846 goes directly over Shap Fell culminating in a steep gradient of one in seventy-five for 4 miles (6·4 km.) at the summit, instead of going the extra 30 miles (48 km.) round the Cumberland coast advocated by George Stevenson. The maximum gradient that Robert Stevenson had on the London to Birmingham railway was 1 in 330, but this was obtained by the high capital cost of the spectacular civil engineering at £50,000 per mile. Joseph Locke would have done it differently; the continuation line from Birmingham to Warrington which he engineered between 1834 and 1837, cost £19,000 per mile, because the only remarkable feature on the line was the Dutton Viaduct not far from Northwich. It is 60 feet (18 m.) high and has twenty arches of 60-foot (18-m.)

Joseph Locke's Dutton Viaduct on the Grand Junction Railway was completed in 1837.

span. Another reason for his lack of fame now is that he was as near faultless as an engineer can be, and that does not commend itself to public notice. His estimates of cost were nearly always correct, whereas Stevenson's and Brunel's could usually be doubled. He was the pioneer of the detailed unambiguous job specification which lent itself to the employment of a single large contractor instead of many little ones which was then the current practice. In this modern style, he and Thomas Brassey built railways all over Europe.

With the opening of the Liverpool and Manchester Railway in 1830, the three essential elements for a modern railway were brought together for the first time: an easily graded track, reliable and consistent locomotion, and the running of the railway under the sole control of one company. Bringing these elements together is the major reason for George Stevenson's deserved fame, but he was nearly fifty and had to back down to the new professionally trained engineer. The line was an immediate success, following which work began on all the other main lines. Robert Stevenson built, among others, the London and Birmingham Railway, with a nearly level gradient to get full and optimum use of steam locomotion. This therefore entailed many heavy civil engineering works, the more troublesome of which he had to take over from the small contractors. These included on the Birmingham line the cuttings at Tring and Blisworth, the immense embankment at Wolverton and the biggest job of them all, Kilsby Tunnel, which ultimately cost three times the original estimate of £100,000 and was the last work to be completed on the 112-mile (180-km.) line. Robert Stevenson calculated that he walked this distance fifteen times in the course of its construction. He would have also travelled it by horse, coach and railway wagon.

The making of Wolverton Embankment in Buckinghamshire was exciting. To bring spoil for the embankment required building a bridge over the Grand Junction Canal, whose traffic the railway would eventually take. The canal company were not inclined to co-operate, so Stevenson with a large number of men moved into the site on the evening of 23 December and worked all night by torchlight and over Christmas until the temporary bridge was completed. On 30 December, the canal company's engineer moved in with his men and pulled the bridge down. Stevenson then won an injunction from the Chancery Court to restrain the canal company from damaging any of the works of the railway company, and so the bridge and the embankment were built. Then the embankment caught fire. Some suspected the canal company of more cloak and dagger action, but it transpired that the bank contained minerals which spontaneously ignited and they burnt themselves out to no ill effect.

Building Wolverton Embankment on the London and Birmingham Railway involved Robert Stevenson in some clandestine night work.

Men at work on the barrow runs in Tring Cutting. Their work in the 2½-mile (4 km.) long cutting for the London and Birmingham Railway was finished in 1838.

Tring Cutting, just north of Berkhamsted, Hertfordshire, was excavated through chalk for $2\frac{1}{2}$ miles (4 km.) and is 57 feet (17 m.) deep for a quarter of a mile (400 m.). Nearly all the work was done by pick and shovel, the only power used was the horsepower to pull the wheelbarrows up the sides of the cutting on planks. A barrow containing a ton or more of spoil was guided up one of the thirty or so barrow runs by a man. If the horse pulled smoothly and the man did not slip on a wet plank all was well and the barrow was emptied at the top. The knack if something went wrong was for the man to fall to one side and push the barrow to the other side. This art must have been well mastered because only one man was killed. Apart from the size of the cutting, there were no difficulties in its construction. At Blisworth Cutting as the line nears Northampton, the rock was more rotten than expected and there were many springs of water. Pumps were brought in, the cutting made wider than planned and retaining walls had to be built in places to support the crumbling rock. Blisworth Cutting is $1\frac{1}{2}$ miles (2·4 km.) long and 65 feet (20 m.) deep.

The hills between Leicestershire and Oxfordshire narrow at the Watford Gap. It is an obvious place for London to Midlands transport to pass through, and road, rail and canal do so. This was the site for Kilsby Tunnel. It took four years to build, and was the last and most difficult work completed on the London and Birmingham Railway. At 2,425 yards (2,278 m.), nearly $1\frac{1}{2}$ miles (2·3 km.), it was of unprecedented dimensions with its provision for two tracks of railway. Here Stevenson was almost beaten by unexpectedly large amounts of quicksand. It was already known that there was quicksand in the ridge because in 1809, when the Grand Junction Canal Company planned to drive a tunnel there, the trial borings showed its presence and the site of the tunnel was moved to the east to avoid it. Stevenson moved his tunnel to the west for the same reason, and although his trial borings revealed some quicksand, they did not show its vast quantity.

The usual method to drive a tunnel is to work from each end and from the bottom of shafts sunk in the hill, so that work can proceed on many faces. Sixteen working shafts were planned for Kilsby, some more than 100 feet (30 m.) deep. At 35 feet (10 m.) shaft No. 2 hit sand and water, No. 3 struck water at 71 feet (22 m.) and No. 6 was drowned out at the same depth. An inclined tunnel called a driftway was put into the hill in an attempt to drain the site by gravity, but it was blocked many times and the idea was abandoned. Then the contractor died of a broken heart and Stevenson was obliged to take over direct supervision himself.

Stevenson moved in on the quicksand at Kilsby with 1,250 men and 200 horses. He sank new shafts parallel to some of the working shafts, made cross links and installed pumps which were set to work. They appeared to make no impression so more were brought in. Ultimately there were thirteen steam pumps pumping out quicksand and water at 1,800 gallons per minute (8,180 litres per minute) for a year and a half before the battle was won and work on the tunnel faces could resume.

From the bottom of the sixteen shafts the men tunnelled in two directions working night and day to make up the time lost to the quicksand. In a game of "Dare" (probably drunken), three navvies lost their lives attempting to jump across the mouths of these shafts. Two of the shafts were not filled in afterwards but left for ventilation of the completed tunnel. In June 1838 the last brick of the lining was laid with ceremony and, led by a band, the company marched through the length of the tunnel to speeches and a feast. The village of Kilsby was probably glad to return to normal after four years of the energies of more than a thousand navvies.

Steam-powered winding gear at the top of a shaft above Kilsby Tunnel.

At the bottom of a working shaft in Kilsby Tunnel. There were sixteen of these shafts and some more for pumping.

Sonning Cutting on the Great Western Railway from London to Bristol gave Isambard Kingdom Brunel a great deal of trouble.

While Stevenson was building the London to Birmingham Railway Isambard Kingdom Brunel was at work on his Great Western Railway running 119 miles (192 km.) from London to Bristol: "The finest railway in England, but not the cheapest," he said. It was built on a 7-foot (2·13-m.) gauge and in a grand style. Stevenson was just building a railway, a difficult engineering job, but Brunel knew he was making something great, a work of art even, and he did not spare the cost. The two works on this line which gave the most trouble were Box Tunnel and Sonning Cutting.

Sonning Cutting was finished by 1840 so that trains could run as far as Reading, but bad weather and its immense size had made it a very difficult task. The cutting is 2 miles (3·2 km.) long and 60 feet (18 m.) deep in parts, and towards the end of 1839 work was stopped by a storm which flooded the Thames and Avon valleys. One thousand two hundred and twenty men and 196 horses were wallowing in the deep mud of the partially completed cutting.

Box Tunnel was the last work to be completed on the Great Western Railway. It is 786 yards (659 m.) longer than Kilsby and measures nearly 2 miles (2,937 m.). Late in 1836 the eight working shafts were put down to the tunnel level, up to 300 feet (91 m.) deep and 28 feet (8·5 m.) in diameter. Two contracts were let: one for the half mile (800 m.) through rock which was unlined, its shape being that of a gothic arch, and one for the remainder which was lined with thirty million bricks. The miners met water in the rock and one inundation flooded the works and filled up a shaft to 56 feet (17 m.). The tunnel consumed a ton (1,000 kg.) of gunpowder and a ton (1,000 kg.) of candles a week, and about a hundred men were killed. A 89 quarter of a million tons (250 million kg.) of spoil were removed and, except for the gunpowder blasts, all the work was done by hand. Ultimately, when Brunel was putting everything in to catch up for lost time, there were 4,000 men and 300 horses on the site plus another 100 horses to move the bricks from a nearby brickyard. Box Tunnel was finished in June 1841. It rises on an incline of 1 in 100 to the east and some authorities say that when the sun rises on 9 April, Brunel's birthday, it shines straight down the tunnel.

Joseph Locke did not like tunnels and he avoided them whenever possible, but he could not avoid the Woodhead Tunnel because the line between Manchester and Sheffield had already been laid out by Charles Vignoles whom he replaced in 1839, two months before work was due to start on the tunnel. But even Joseph Locke could not have gone through the Pennines without a tunnel, although he would have tried and maybe the tunnel would have been shorter and in a different place. The tunnel was to be 3 miles 22 yards (4·85 km.) long, the longest of its time, but only for a single track because the company was short of money. It was a tremendous undertaking and George Stevenson said he would eat the first locomotive to pass through.

The portals of Woodhead Tunnels seen from Woodhead Station in 1903. The first tunnel is on the right.

Locke looked over the specification and immediately doubled Vignoles' estimate to £200,000. The accommodation of forty stone huts for men and horses, and 4 miles (6·4 km.) of service road had already been made, and by the end of 1839 the five shafts 8 feet (2·4 m.) in diameter had been put down to the tunnel level, the deepest of which was 567 feet (173 m.). The tunnel was built by mining methods using 157 tons (160,000 kg.) of gunpowder on a 1 : 200 gradient rising eastwards. Three-quarters of a million tons (750 million kg.) of spoil were removed, half of it up the shafts. Woodhead was a desolate, remote place then and everything was appalling: the rock, the company, the men. The rock was unstable, and waterlogged in places. Pomfret, a surgeon employed by the men themselves, said that they were a drunken and dissolute lot, whose wives were swapped and sold. The attitude of the Company left a lot to be desired too. For example: the holes for the gunpowder were hand drilled. The charge was then inserted and rammed home with an iron stemmer. On one occasion William Jackson, miner, was looking over the shoulder of John Webb who was stemming. The iron caught the rock, sparked and ignited the powder. The stemmer was fired straight through Jackson's head. Purdon, the resident engineer, refused to use copper stemmers which would not spark because they slowed the work. Thirty-two men were killed there which is a bigger proportion of the total than at the Battle of Waterloo. The tunnel was inspected for the Board of Trade by General Pasley in December 1845, who said it was the best tunnel he had ever seen. If Locke had to build tunnels, he did them well.

However, the Woodhead Tunnel soon became a bottleneck for traffic, so in 1847 work was started on a second tunnel parallel and of the same dimensions (namely 15 feet (4·6 m.) wide, 18 feet (5·5 m.) high with a semi-elliptical arch). The working conditions were better this time and the job was finished in 1852. In the original tunnel twenty-five arches were put in the north wall and the second tunnel was built from them. Work was held up in 1849 when cholera, which was rife over much of England at the time, struck the camp and twenty-eight men died. *The Manchester Guardian* at the time said they died of imprudence and intemperance. The tunnels were finally abandoned in 1954 because maintenance could not keep pace with their deterioration, and they now carry cables for the National Grid. A description of the new Woodhead Tunnel is included in Chapter 9.

A bridge can be appreciated in one view. It exists as an object, unlike a railway which is large and nebulous, or a tunnel which is invisible. It is therefore easier to understand the engineering genius of Brunel and Stevenson by looking at some of their bridges.

The Britannia Bridge over the Menai Straits between mainland Wales and Angelsey, was completed in 1849 with two centre spans of 460 feet (140 m.). The idea of using one of the roadways on Telford's suspension bridge a mile (1·6 km.) away to the northeast was rejected because suspension bridges are flexible and unsuitable for the heavy loads that a railway imposes. So Robert Stevenson set out to design a rigid suspension bridge by carrying the double railway in two stiff rectangular tubes. A model was built by William Fairbairn and tested to destruction, which revealed that the suspension chains were unnecessary.

The Conway Bridge on the same Chester to Holyhead Railway as the Britannia Bridge is also tubular, but smaller. It was therefore very suitable for practising the method of erection. This was to build the tubes at the edge of the water and float them into position by placing pontoons under each end. Nothing unforeseen occurred at Conway except that Edwin Clark, resident engineer for both bridges, severed his big toe in a capstan. They were now ready to tackle the big job on the treacherous Menai Straits.

Robert Stevenson's Britannia Bridge over the Menai Straits. Two 1,511-foot (460 m.) long tubes carried the rails 100 feet (30 m.) above the water.

The site chosen by Stevenson for his Britannia Bridge was to the west of Telford's Menai Straits Bridge where a convenient rock in midstream gave its name to the bridge. Work on the masonry piers was started in the spring of 1846. The two end pairs of tubes were constructed *in situ* on scaffolding and the central pairs of tubes, each weighing 1,500 tons ($1\frac{1}{2}$ million kg.), were built on staging by the bank. A thousand yards (900 m.) of timber staging were erected and work started on the wrought iron tubes in June 1847. Some of the plates were $3\frac{1}{2}$ inches (8·9 cm.) thick and although they had been passed through rollers in their manufacture, they had to be flattened by men swinging 40-pound (18-kg.) hammers, which made an awful din. Each one of the 900 tons (914,000 kg.) of rivets was hand closed. Edwin Clark described the scene at night as an interesting sight, when red-hot rivets were thrown from the forty-eight hearths over 30 feet (9 m.) by men and boys using tongs. The staging partially collapsed as the weight of the tubes increased, and the 9-inch (23-cm.) upward camber at the centre of the tubes had to be restored by banging in wedges underneath.

Two of the four larger central span tubes for the Britannia Bridge being assembled on wooden staging on the shore of the Menai Straits.

Floating the second Britannia tube on 5 November 1848.

The first of the tubes was floated on 19 June 1848. There were grandstands on the three remaining tubes and thousands of people assembled to watch. Bands played and cannons were at the ready to be fired at the moment of success. Robert Stevenson was in sole charge, but Brunel and Locke were there to give moral support. Pontoons were put under the tube at low tide and were floated on the high tide at 6 p.m., but a capstan broke and the work was closed for the day. The next day was very windy and the boats had considerable trouble, but nevertheless the floating tube was released at 7.30 p.m. on the high tide. It was caught by wind and water, an 8-inch (20-cm.) cable broke and it looked as if 472 feet (144 m.) and 1,500 tons ($1\frac{1}{2}$ million kg.) of iron were about to be lost down the Menai Straits. A 12-inch (30-cm.) cable capstan was pulled from its anchorage as the tube slewed round. The foreman in charge of this capstan, who had been knocked into the water, swam ashore and invited the crowd to pull on the cable. They did, in their Sunday clothes. Hundreds of people were dragged towards the water by the momentum of the tube, they held and finally hauled in. The capstans at the other end then swung it round across the stream and as the tide fell, the tube was left safe and dry on ledges prepared for it at the bottoms of the piers. The bands played victorious music and the cannons were fired, and Stevenson went to bed. The pontoons were recovered around midnight downstream, and there were no casualties.

The hydraulic jacks which were to raise the tubes were assembled at the

One of the hydraulic jacks in the Britannia Tower which has nearly completed its
work of raising the tube. The eyebar chains are fixed to the roof of the tube.

Britannia Bridge in 1973 supported by steel arches after the original spans were severely damaged by fire.

top of the towers in June and July and then the lifting began. The jack pushed its piston and connecting rod vertically upwards and with them went the cross-head to which was connected eyebar chains fixed to the roof of the tube. The stroke of the jack was the same as the length of the links in the chain, so that when the jack was at the top of its stroke the tube was wedged, a link taken out of the chains and the jack brought down for another stroke. The slot up which the tube travelled was filled in with masonry as the tube rose, which was just as well because on one occasion the jack broke and 50 tons (50,000 kg.) of cross-head and gear fell on to a man climbing a rope ladder from the tube to the jacking platform. He was killed, of course, but the tube only fell a few inches to solid masonry instead of to the bottom of the Straits. The first tube was at full height on 13 October 1848. When the jack was being lowered from the tower a sailor was crushed by being dragged round a capstan by the rope he was controlling, but the jack was recovered intact. The three other tubes were floated and raised with minor difficulties. When in position, they were joined through each tower to make two 1,511-foot (460·6-m.) long tubes. Thus, the strength of what were simple beams was increased by employing the cantilever principle as well. (See page 109 for an explanation of cantilevers.) The long composite tubes were fixed to the central tower but free to move under the effects of heat expansion through the other towers on 264 cast iron rollers and 132 gunmetal balls.

On 18 March 1849 a 500-ton (500,000-kg.) train ran through one of the tubes and on 19 October both lines were open. The bridge was severely damaged by fire in the night of 23 May 1970 and the deck is now supported by steel arches.

The dimensions of the Royal Albert Bridge are similar to the Britannia Bridge and it crosses the River Tamar with two 465-foot (141·7-m.) spans clearing the water by 100 feet (30 m.). But Brunel had to build his central pier in 50 feet (15 m.) of water and take it a further 80 feet (24 m.) down to bedrock for a foundation. The river bed was excavated using a pneumatic caisson, in which air pressure was used to keep the river water out and away from the men who were digging on the river bed. (See page 102 for a further description of pneumatic caissons.) This was the first time one had been used at this depth and on a major bridge. The central pier had been built to above the water level by 1856 and the caisson was removed. Meanwhile work had started on the two main girders at the side of the river. It was intended to float them into position and then jack them up, a repeat of the operation at Britannia Rock except that the piers were built as the jacking operation proceeded.

The day for the floating was 1 September 1857. Docks had been cut in

The Royal Albert Bridge under construction. One girder has been jacked up to its final position and the other is in its lower position waiting to be raised.

the river bank so that the pontoons could be slid under the trusses at low tide. The usual crowds and bands were there and the weather was perfect. Brunel had worked out a system of communication with his winches and capstans which involved numbered boards and coloured flags. The tide rose and 1,000 tons (1 million kg.) of girder was lifted. The whole area fell silent as Brunel had requested and he stepped out on to a control platform on the girder. The signal flags were waved and the number boards displayed. The girder floated out, turned and deposited itself on the stubs of the piers as if guided by one man; it had been. The bands played and the church bells rang and the crowd cheered. The girders were jacked up and the masonry of the piers built under them, until they were at their full height. The bridge was opened by Prince Albert in May 1859, but Brunel was not there to share the triumph; he was dying, broken by overwork on his big ship the *Great Eastern*. He had a preview of his bridge while lying paralysed on a railway flat-truck which carried him slowly across.

Both the Albert and the Britannia Bridges were special answers to special problems. An arch could not be used because the Admiralty wanted the whole width of the streams to be unimpeded, and the material for a big cantilever, steel, was not available yet. Robert Stevenson had invented a new type of bridge in his Britannia Tubular Bridge but only a few were built. He was also responsible for the development of the bowstring girder bridge and many more of these were built. The Albert Bridge is of this type, but it is better typified by his High Level Bridge at Newcastle. The outward thrust of the arch member is taken by a long "bowstring" tie of wrought iron joining the ends of the arch.

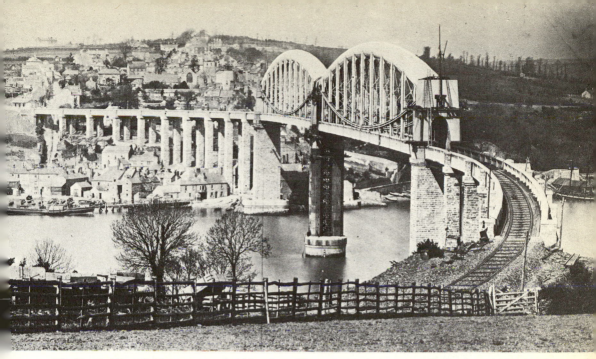

The Royal Albert Bridge was completed in 1859. The two central spans are 465 feet (142 m.) each and 100 feet (30 m.) above the water.

At Newcastle-upon-Tyne, there are six bowstring spans of 125 feet (38 m.), and each is made of four ribs. The railway is carried on the top deck over the arches and the roadway is on the lower deck suspended from the arches. On 6 October 1846, the first pile for the foundations was driven by a steam hammer, used here for the first time for this purpose. The pile went in 32 feet (9·7 m.) in four minutes. There was no trouble in the construction apart from the middle pier. Here, quicksand kept flowing into the coffer dam and prevented the masons building the pier on the pile heads. Huge quantities of chalk were put around but it was no good. Finally, the coffer dam was filled with concrete to the heads of the piles. Then the water could be pumped out and the masonry pier built. While working on the girders, a man stepped on to a loose board and fell; not very far though, because a nail in the planks of the centring caught his trousers and he hung there, well over 100 feet (30 m.) above the river. His tailor used this incident as a testimonial. The bridge was opened in August 1849.

Joseph Locke, Robert Stevenson and Isambard Brunel, all died within a year of one another at the end of 1859 and the beginning of 1860, in their middle age and worn out. Their deaths marked a turning point in civil engineering. They caused the change by their works. They were public heroes, responsible for the apparent triumph of man over nature. They were seen to lose occasionally and this made the battle interesting, not clear cut. This glory was not for the engineers that followed. The new ground had

Robert Stevenson's High Level Bridge at Newcastle. There are six bowstring girder spans of 125 feet (38 m.) each. The bridge was opened in 1849, the same year as his Britannia Bridge.

been broken. As engineering became more complex and scientific, the public could not appreciate what was being achieved, and besides, there was not one single engineer for a job but a team. The large contractors are now the well-known names, but they are not heroes. The glorious days of civil engineering ended in 1860.

8

Railway Engineering:
Big Bridges

The railway engineer had to face problems in bridge building that had not occurred to the highway engineer. The road-maker can easily divert his route to enable him to take advantage of the most convenient point for a river crossing, that is where the river is narrow or the ground is good for the foundations. The railway engineer not only has to maintain the direction and the level of his route, but also his bridges have to carry much bigger loads. This challenge was met by fresh ideas and new materials. Cast iron replaced stone, wrought iron replaced wood, and the new thinking can be seen in Robert Stevenson's Britannia and High Level Bridges already described. Then steel superseded iron and the railway engineer could really build some awesome bridges. The climax was reached by Benjamin Baker with his Forth Railway Bridge in 1890. The best forms for a long span railway bridge are the cantilever and the arch because they are not likely to deform under concentrated loads. For shorter spans a simple girder bridge will do. The first notable steel arch was James Eads' St Louis Bridge of 1874.

Before the Civil War, James Eads (1820–87) had walked on the bed of the Mississippi in a diving bell 65 feet (20 m.) below the flood waters in connection with his salvaging operations. He had seen the power of the river, had seen the sand flow and the effects of ice on the bottom of the channel. The Mississippi is treacherous and unpredictable. A storm can raise its level by 40 feet (12 m.) and increase its flow from 3 to 9 miles (5 to 15 km.) per hour. It freezes once a year and the ice is horrific when the rising waters from the Missouri rivers sweep down the valley before the ice has properly melted. When Eads designed his three steel spans of 500 feet (152·4 m.) each, rising 50 feet (15 m.) above the highest level that the river had ever reached, he knew that the piers for his bridge would have to go right down to solid rock. Pile foundations would have been carried away in no time. There had to be a bridge at St Louis because the river which had brought prosperity to the city was now an obstacle to the railways, the new bringers of prosperity. It was certain that the city would be by-passed without this

James Eads' three 500-foot (152 m.) steel spans across the Mississippi at St Louis carried a road on the top deck and railway tracks on the lower level.

bridge. After considerable opposition by the vested interests of the rivermen, in 1869 the western shore abutment was built in a coffer dam down 30 feet (9 m.) to bedrock. Eads then changed his mind about coffer dams and used pneumatic caissons for the three other piers. Just as well, because the west river pier goes down 86 feet (26 m.), the east river pier 122 feet (37 m.) and the east shore abutment 135 feet (41 m.).

The method with the caissons was to float them into position, sink them to the sandy bottom of the river and surround them with ice-breakers and other protections. The masonry of the pier was then built on and inside the caisson as the men working in the compressed air at its base excavated the material and passed it through the air locks to the surface. The air in the working chamber was equal to the pressure of the water in the sand outside. In this manner the water was kept out and the piers were eventually founded on solid bedrock, taking between six months and a year to construct. Twelve men died of caisson disease in the east river pier, and Eads called in his family doctor who worked out basic rules for men working at such high pressures so that never again would so many men die from this cause. Caisson disease is the same as the divers' complaint called the "bends", and is due to nitrogen in the air dissolving in the blood under pressure and being released as bubbles when the pressure returns to normal. Its symptoms are cramp and paralysis. The bubbles can be trapped at the joints in a man's body causing cramp and in his heart causing death. Decompression was slowed down and the men were forced to rest for thirty minutes after coming out and were fed warm food. Alcohol was forbidden. Men are exhilarated by the extra oxygen under pressure, and work and sweat profusely. Voices sound thin

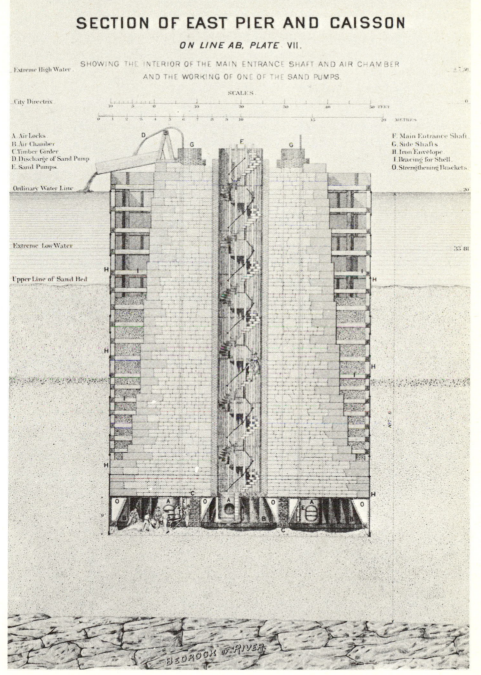

SECTION OF EAST PIER AND CAISSON

ON LINE AB, PLATE VII.

SHOWING THE INTERIOR OF THE MAIN ENTRANCE SHAFT AND AIR CHAMBER
AND THE WORKING OF ONE OF THE SAND PUMPS.

SCALES.

Extreme High Water.

City Directrix

A. Air Locks
B. Air Chamber
C. Timber Girder
D. Discharge of Sand Pump
E. Sand Pumps.

F. Main Entrance Shaft.
G. Side Shafts
H. Iron Envelope
I. Bracing for Shell
O. Strengthening Brackets.

Ordinary Water Line

Extreme Low Water

Upper Line of Sand Bed

BEDROCK OF RIVER

Cross-section of a pneumatic caisson used on the St Louis Bridge foundations.

103

and cracked, and the senses of smell and touch diminish. There is also a fire risk from the extra oxygen. As the depth and the pressure increased so the working period was reduced until after 100 feet (30 m.) there were two working periods of 45 minutes each per day. Only one man died during the construction of the 135-foot (41-m.) deep east shore abutment and he had been drinking beer.

In spite of these precautions taken to protect men working in compressed air, there is virtually no avoiding action which can be taken to guard against a pneumatic caisson suddenly diving into softer ground. Such an accident occurred in St Petersburg in 1876 two years after the St Louis Bridge was finished. Twenty-eight men were working on the soft bed of the River Neva when the pneumatic caisson they were in plunged into a pocket of softer material. Nine managed to get up the shaft as mud and water went into the working chamber but the rest were trapped. It took twenty-eight hours to reach them and by then only two were alive. The caisson was cleared and work resumed. A year later the air lock on the shaft gave way and nine men were blow up the shaft by the escaping air and killed. Twenty others in the working chamber were smothered by mud and water. Fortunately for Eads the bed of the Mississippi was consistent and his air locks did not fail, so these extreme troubles were avoided.

Erection of the west arch of the St Louis Bridge using temporary cantilevers.

The arches of the St Louis Bridge are complete in this picture and are self-supporting; construction of the deck has begun.

Once the abutments and piers of the St Louis Bridge were in place and founded on bedrock, work could proceed on the steel arches. They were erected as cantilevers, one half of one holding up and balancing another until two halves met at the middle. When the arch was complete, the temporary cantilevering could be removed. The more usual falsework for an arch could not be used because of the nature of the river already described. Each arch is made of four ribs transversely braced by wrought iron. Each rib is two 18-inch (46-cm.) diameter tubes, one 12 feet (3·7 m.) above the other. The tubes are made of six chrome steel staves 12 feet (3·7 m.) long and carefully machined to fit together like a barrel. There was the usual tense moment when the last piece of the arch came to be fitted, and as usual it did not fit, because such large structures as this are noticeably influenced by temperature changes. (Eads had calculated that the final bridge would rise and fall at each crown by 8 inches (20 cm.) with the temperature variations.) The piece in question was too big so the erected tubes were packed with ice to contract them but still they did not fit. Eads had to design a special segment which could be put in place and then extended by screws.

The bridge has two decks and was opened on 4 July 1874. As a party piece fourteen locomotives went across two abreast and then in a single unbroken line. Eads put his roadway over the tops of the arches at St Louis. The alternative is to suspend the deck under the arches and this was the   105 shape chosen by Gustav Lindenthal in 1916 for his Hell Gate Bridge, and by Ralph Freeman for the Sydney Harbour Bridge.

The Hell Gate Bridge, New York, crosses the East River near its junction with Long Island Sound. It has a span of 977 feet (298 m.), the longest arch in the world at the time and for the next sixteen years. The arch was erected as two cantilevers, just as Eads had done at St Louis, and the heavier sections were held together by the biggest ever rivets: 4 inches (10 cm.) in diameter by 11 inches (28 cm.) long. It is still the most heavily loaded bridge in the world carrying a live load on its four railway tracks of 22,000 pounds per foot (32,400 kg. per m.) of its length. (The strongest suspension bridge is the George Washington Bridge capable of carrying 5,000 pounds per foot (7,560 kg. per m.).)

Sydney Harbour Bridge, completed in 1932, has a steel arch span of 1,650 feet (503 m.).

The longest steel arch span is O. H. Ammann's Bayonne Bridge, New York, but the best of them is Sydney Harbour Bridge. It is only 25 inches (64 cm.) shorter and can carry a load of 12,000 pounds per foot (16,200 kg. per m.) of its length, which is nearly twice that the Bayonne Bridge can carry. Sydney Harbour Bridge has four railway tracks, a 57-foot (17-m.) wide roadway and footpaths on a 160-foot (49-m.) wide deck which goes through and is suspended from the 1,650-foot (503-m.) arch. Ralph Freeman was its designer and Dorman Long were the contractors.

Lawrence Ennis, in charge of the erection, started on the foundations in January 1925. These were quite straightforward, because there was good rock only 30 feet (9 m.) down. They were made of solid concrete 90 feet (27 m.) by 40 feet (12 m.) and 30–40 feet (9–12 m.) deep. The four bearings on the two blocks were to take the arch thrust of 19,700 tons (20 million kg.) at 45°. The contractors erected two new and well-equipped workshops on the north side of the harbour to prefabricate the steel; the proximity of these meant that the parts for the bridge could be larger and heavier than otherwise because there would be no transport problems. Eight hundred men working there in three shifts treated all the 50,000 tons (50 million kg.) of steel, 37,000 tons (37 million kg.) of it for the main arch. They handled steel angles up to 12 inches (30 cm.) by 12 inches (30 cm.) and $1\frac{1}{4}$ inches (3·2 cm.) thick, and the thickest plates were $2\frac{1}{8}$ inches (5·5 cm.) thick. Two hundred and fifty men opened up a new quarry on the coast 150 miles (240 km.) south of Sydney and worked there for six years, supplying crushed and dressed granite which was taken to the sea in three specially constructed

The cantilevered half-arches of Sydney Harbour Bridge. Sixty miles (96 km.) of $2\frac{3}{4}$-inch (7-cm.) diameter cables tied to the ends of the upper chords held them up.

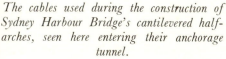

The cables used during the construction of Sydney Harbour Bridge's cantilevered half-arches, seen here entering their anchorage tunnel.

vessels. The most difficult and spectacular part of the contract was the building of the arch itself. Lawrence Ennis was ready to start on this in October 1928. It could not be built upon centring, as the Bayonne Bridge was, because the deep and wide channel prevented any support from underneath. The arch was therefore built out from each side as two cantilevers, in the same way that Eads had done at St Louis except that at Sydney Harbour, the cantilevers were tied back from the ends of the upper chords by 128 cables. Each $2\frac{3}{4}$-inch (7-cm.) cable was fixed to the top of an end post, passed round a u-shaped tunnel cut 100 feet (30 m.) into the rock and back to the top of the other end post.

Ralph Freeman had designed two special cranes to run along the top chord of the arch, building as it went. Each crane weighed 565 tons (574,000 kg.) and could lift 122 tons (124,000 kg.). The parts of the arch were brought on pontoons from the workshops to underneath the end of each cantilever and lifted by the cranes. Once the new parts had been riveted into their places, the cranes hauled themselves farther up and along to repeat their job. The biggest rivets were $1\frac{3}{8}$ inches (3·5 cm.) in diameter and 12 inches (30 cm.) long. On 14 August 1930, the cantilevers were complete and ready for joining to make the arch. To accomplish this, they were lowered by letting out the cables holding them up. One cable would be slackened off a little, which threw a bigger load on all the other cables which would stretch by a small amount, and the two ends of the cantilevered half-arches would move very slightly closer. Ennis let the cables out 3 inches (8 cm.), then 4 inches (10 cm.) until at four o'clock in the afternoon of 19 August, the two cantilevers just met. Then the temperature dropped causing them to contract and they moved apart again. The men on the cables could not keep pace with the contraction at first and it was not until ten in the evening before the 8-inch (20-cm.) pins joined the two halves. Now the cables were removed and the arch existed for the first time. The suspenders and the deck were erected next from the middle as the creeper cranes crept back down the arch. A man fell 172 feet (52 m.) into the water at this stage and broke two ribs but he was back at work after a couple of weeks, and all the steelwork was finished in May 1931. Detail work occupied nearly a year and Dorman Long handed over the finished bridge to the New South Wales government in March 1932. As a test piece, seventy-two locomotives weighing 7,600 tons (8 million kg.) steamed across four abreast.

The 285-foot (87-m.) pylons at each end are for decoration and serve no engineering function. They were made of concrete faced with granite. Some of the 60-miles (96-km.) of cable used to hold up the half-arches when they were cantilevers, were used in 1936 to make the 600-foot (183-m.) span Indooroopilly Suspension Bridge, Brisbane, which is still the longest in Australia.

Benjamin Baker (1840–1907) decided that the cantilever would be a better form for the two 1,710-foot (521-m.) spans for the Forth Railway Bridge over the 200-foot (61-m.) deep Firth of Forth in Scotland. A cantilever is a horizontal beam held at one end only. It is usually balanced by the weight of a similar beam rigidly fixed to it on the other side of that end, so that in effect a cantilever bridge consists of a series of fixed see-saws. There are three such see-saws in the Forth Bridge, joined by two suspended girders and approached by normal girder viaducts. At the site chosen by Baker the firth narrows to a mile (1·6 km.) wide between the Queensferry and Fife shores, and in midstream there is the Islet of Inchgarvie.

The contractor, William Arrol, started work on the foundations in January 1883. The piers for the Fife cantilever and viaduct were built upon virtually dry land and gave no trouble with their foundations. Those for the Queensferry viaduct were in shallow water and built in coffer dams. A jetty of 2,000 feet (610 m.) was thrown out to the Queensferry cantilever foundations at the edge of deep water. Some of the foundations for the Inchgarvie and Queensferry piers were too deep for coffer dams to be used,

The Forth Railway Bridge just after its completion in 1890. Its two 1,710-foot (521-m.) spans were the longest in the world.

and pneumatic caissons were employed. Each cantilever is supported by four 70-foot (21-m.) diameter piers.

The six caissons required were built in a small bay to the east of the bridge on cradles so that they could be slid into the water. Then they were towed into position and sunk, to rest on the river bed by filling the double skin with concrete. Each caisson was 70 feet (21 m.) in diameter, made of wrought iron with a steel cutting edge at its lower rim. The first caisson was launched on 26 May 1884, attended by a band and the Lord High Commissioner and the Countess of Aberdeen. A hydraulic ram set it in motion down the slipway and it glided majestically into the water, tilted at an alarming angle while it explosively relieved itself of the air trapped in the 7-foot (2·1-m.) high working chamber beneath its base, and then it rocked to an upright position. When the caissons had reached their foundation depth, at a maximum of 90 feet (27 m.) below the water level, the bedrock was levelled and the caissons completely filled with concrete, the incoming material having to pass through the air locks. Due to close medical supervision, there were no deaths from the "bends".

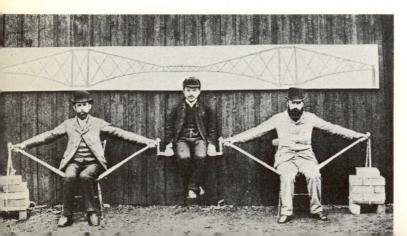

Human model of the Forth Bridge. The central figure is Mr Kaichi Watanabe, a student with Baker, who subsequently became Chief Engineer to several Japanese railway companies.

One of the pneumatic caissons for the Forth Bridge on its slipway.

Above the low water line the piers were faced with granite blocks and they rise to 18 feet (5·5 m.) above the high water mark. Then there are two bed-plates, the lower is fixed firmly enough by forty-eight 2½-inch (6·4-cm.) diameter bolts which extend 24 feet (7·3 m.) into the pier to anchor plates. The upper bedplate is fitted to the lower in such a way that it may move slightly as the cantilever it supports moves under the effects of heat. All the caissons, except one, were sunk without mishap. The unlucky one was the northwestern of the Queensferry group. Its double skin had been partially filled with concrete when work stopped for the New Year's holiday. It was constrained by hawsers but allowed to rise and fall with the tide, resting on the bottom at low tide. On Sunday, 1 January 1885, it stuck in the mud and did not rise with the tide. It filled with water and capsized. It took ten months of work, with the expense of two men's lives, to retrieve this caisson.

All the steel for the cantilevers was prepared on the shore. Each plate was heated in gas ovens and hydraulically pressed to the right curvature. The tubes were then assembled, drilled and numbered before being taken out on to the firth. The plates for the vertical columns were 1¼ inches (3·2 cm.) thick. These 12-foot (3·7-m.) diameter columns were the first part of

Prefabrication of a 12-foot (3·65-m.) diameter tube for the Forth Bridge in 1886.

111

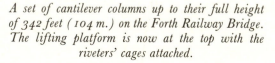

A set of cantilever columns up to their full height of 342 feet (104 m.) on the Forth Railway Bridge. The lifting platform is now at the top with the riveters' cages attached.

the steelwork to be erected, using a special lifting platform which held the columns in position and allowed the distance between them across the width of the bridge to decrease from 120 feet (36·6 m.) at the piers to 33 feet (10 m.) at the top. This batter was echoed throughout the length of the bridge. The riveting was done by hydraulic machines and the men operating them were inside wire mesh cages which moved up with the lifting platforms. The first lift was in October 1886 and the columns reached their full height in April 1887. In building these columns seventeen men were killed which greatly exceeded the yearly average of nine. Sixty-three men were killed in the seven years it took to build the bridge, but this was mainly due, say contemporary accounts, to the men's carelessness. Many injuries were caused by falling rivets. A man standing on top of the columns is 342 feet (104 m.) above the piers. Baker wrote "I saw a hole one inch in diameter made through the four inch timber of the staging by a spanner which fell about 300 feet and took a man's cap off in its course." Two men in a rowing boat were stationed at the base of each cantilever to pick up men who fell. Eight men were saved in this way, and thousands of caps.

To a distant observer the arms of the cantilevers appeared to grow in perfect organic symmetry. All the tension members are open lattice girders and the compression members are tubes. The tubular part of the cantilever springs from the massive joint of five tubes called the skewback which is fixed to the upper bedplate on each pier. There are six tubes in each lower chord of the cantilever, starting with a 12-foot (3·7-m.) diameter tube and diminishing at each stage until it is 3 feet (91 cm.) square. The upper chords of the cantilevers are made of lattice girders, six on each side. The upper and lower chords are joined by six tubular struts and six lattice ties on each side. As the arms extended, so did the internal viaduct which was to carry two tracks of railway line. The cantilevers were finished in June 1889 and there only remained the suspended girders to close the gaps of 346 feet (105 m.) These spans were built piece by piece as extensions of the cantilevers and rigidly fixed to them. Excitement rose as the two halves approached. Baker had calculated that the bolt holes would coincide at 60° Fahrenheit (15 °C).

This sequence of photographs shows the progress made on the erection of the Forth Bridge cantilevers from April 1888 to September 1889.

The sun shone and the engineers waited. They could not really have expected the holes to fall in line, not when the bridge is 1½ miles (2·4 km.) long, and the particular span is a third of a mile (521 m.). Hydraulic rams were brought to bear and when this was not enough, cotton waste and naphtha were lit around the bottoms of the girders to make the metal expand. This did the trick and the bolts went home. Finally, the bolts and wedges holding the girders to the cantilevers were removed, again with the expansive aid of fire. In November 1889 the last few bolts sheared under the intense load and broke with a loud report which echoed throughout the 50,000 tons (50 million kg.) of steel, nearly seven years after William Arrol had moved on to the site. The great bridge was finished and it was opened by the Prince of Wales on 4 March 1890 and Messrs Baker and Arrol became Sir Benjamin and Sir William.

Twenty years before, there had occurred on the same railway line, the most famous of all bridge collapses: the Tay Bridge disaster. Before 1877 the train journey between Dundee and Edinburgh involved two ferry crossings of the Firths of Tay and Forth. Thomas Bouch had dreamed all his life of

The lives of 75 people were lost when the Tay Bridge was blown over as a train crossed it.

building the two bridges so that the journey could be made in one train and much more comfort. After his Tay Bridge was blown over on the last Sabbath day of 1879, his plans for a Forth Suspension Bridge, on which work had already started, were quickly abandoned.

The Firth of Tay is just over 1 mile (1·6 km.) wide at the place of crossing, but the bridge was (and is, because it was rebuilt in 1882–87 by W. H. Barlow, who designed the beautiful St Pancras Station roof) nearly 2 miles (3·2 km.) long because of its oblique crossing and the curves at each end. It was the longest bridge in the world at the time. Bouch had planned a lattice girder bridge comprising eighty-five spans on brick piers founded on rock which was expected not far down all the way across. After a start had been made the rock was not to be found, so Bouch changed his plans to having most of the piers made of slender cast iron columns on a base of brick and concrete supported by pile foundations. The thirteen central spans, eleven of 245 feet (74·7 m.) and two of 227 feet (69·1 m.), were known as the "high girders" because the single railway line ran level with their bottom chords, whereas it ran on the top chords along the rest of the bridge.

Bouch asked Sir George Airey, the Astronomer Royal in London, what wind pressures he should expect in the Firth of Tay. Twelve pounds per square foot (58 kg. per square m.) was the absurdly low answer and Bouch designed accordingly. If blowholes appeared in the iron columns during their casting, and there were many, the foreman in charge, paid on piece-work rates, filled them in with a mixture of his own devising which would prevent the columns being scrapped. His filler consisted mainly of beeswax and boot polish. The man in charge of maintenance after the bridge was opened was inexperienced and unqualified. When he found some ties loose he merely ordered packing to be put in their joints. The Tay Bridge had to fall down: it was the only decent thing it could do. The Committee of Inquiry blamed the faulty design, construction and maintenance of the bridge for its collapse. Thomas Bouch, as engineer, was responsible for all three factors, so he died four months later.

On 28 December 1879 Driver Mitchell had taken his train out of St Fort Station at 7.00 p.m. bound for Dundee and arrived at the south end of the bridge at 7.12 p.m. He stopped and received permission to cross from signalman Barclay, who had to crawl back to his cabin, the wind was so strong. Barclay watched the train go on to the bridge, watched its tail lights get dimmer and watched a flame of sparks from the flanges of the wheels as they rubbed on the sides of the rails, the wind was so strong. The lights and sparks disappeared from his sight and he signalled his colleague at the north end of the bridge. No answer: his telegraph instrument was dead. The moon shone for a moment and Barclay could see that all the high girders were gone. A diver went down as soon as the weather would allow and he found the train inside the girders on the estuary bottom. The wind blowing on the extra area of the train had proved too much for the ill-fated bridge.

It is fitting to close this chapter with the Tay Bridge disaster because it reminds us that engineers are fallible, and that they are not superhuman. They have to pit their wits against many unknown quantities and sometimes they lose. The unfortunate Bouch certainly lost.

9

Railway Engineering: Tunnels

The Alpine tunnels are the climax stemming from the Box, Kilsby and Woodhead Tunnels, described in Chapter 7. They are the culmination of tunnelling methods started in ancient times. Mont Cenis was the first of the long Alpine tunnels and shows the confidence that engineers had gained by the late nineteenth century.

The Mont Cenis Pass is 6,772 feet (2,064 m.) above sea level and is on a direct line between Lyons and Turin. It was the obvious route for a link between the railway systems of France and Italy. The idea had been around since 1840, but no one then would have seriously contemplated a tunnel of $7\frac{1}{2}$ miles (12 km.) with a mile of rock above it, as there could be no working from shafts. Even the first Alpine tunnel on the Semmering Railway took six years (1848–54) to build and it was less than a mile (1·6 km.) long. However, the Mont Cenis Railway had to be built because Italy was cut off from the rest of industrializing Europe. The Italian parliament passed the authorizing bill in 1857 and King Victor Emmanuel set off the first charge of black powder at the Bardonnechia end in Italy. This was the first of over three million such detonations in the next fourteen years.

The engineer was Germain Sommeiller who had been carrying out preliminary surveys. A lot of work had to be done before a start could be made on the tunnel proper: setting up workshops and accommodation, and making an accurate survey of the mountains above to get the line of the tunnel correct. A slow start was made on the tunnel because all the drilling was done by hand. After three years a third of a mile (536 m.) was pierced at the French end and half a mile (804 m.) in Italy, and at this rate the tunnel would take thirty years to complete. At that time coal miners in England were using a steam drill patented by Thomas Bartlett which Germain Sommeiller improved and adapted to use compressed air. In 1861 this pneumatic drill was introduced to the tunnel, which then progressed three and a half times faster. Water-power plant was installed at each end to compress the air which was taken to the drills in an 8-inch (20-cm.) pipe along the tunnel. It also improved the ventilation in the tunnel. For another ten years men worked in the mountain removing over a million cubic yards of rock. No unexpected troubles were encountered and on Christmas Day 1870 the pilot headings met, the French being two feet (60 cm.) higher than the Italian. By the summer following, the Mont Cenis Tunnel was finished. Now that Germain Sommeiller had pioneered the way, others followed, culminating in the 12¼-mile (19·7-km.) Simplon Tunnel at the end of the century.

The next inevitable step was for a tunnel to be built under a river and the Severn Tunnel was the first. The River Severn is wide and deep, and the Great Western Railway Company crossed it by a ferry at a place called New Passage. The ferry was inconvenient and unreliable because the Severn has a 50-foot (15-m.) tide at times, and the currents are swift causing the mud banks to change their positions. A tunnel had to be built.

In 1871 Charles Richardson drew up plans for a site near to the ferry. The river here is 2 miles (3·2 km.) wide but the tunnel had to be 4½ miles (7·2 km.) long to maintain a reasonable gradient of 1 in 100 and to get under the deep channel of the river near the Welsh side called the Shoots. The water is 95 feet (29 m.) deep at spring high tide. In 1873 a 15-foot (4·6-m.) diameter shaft was sunk 200 feet (61 m.) deep on the Welsh side. It is known as Old Shaft and goes below the tunnel level to act as a drain. An almost horizontal pilot heading was driven out from the bottom of the Old Shaft under the Shoots and on towards England. Headings were also driven from the Old Shaft on the line of the tunnel in both directions, and from other shafts: Sea Wall, Marsh, and Hill Shafts. A second shaft, called Iron Shaft, was sunk at the side of the Old Shaft.

This was the state of affairs on 18 October 1879, the day when the "Great Spring" broke in. The engineers, who were fearing water breaking in under the river, were surprised to find that the water trouble came in the landward heading from the Old Shaft. They had tapped a huge source of fresh water

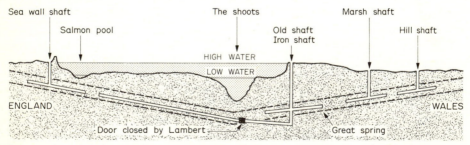

Cross-section of the Severn Tunnel workings on 18 October 1879, the day when the Great Spring broke in.

and it was to take fifteen months to bring the spring under control. All the workings in contact with the Old Shaft were filled within twenty-four hours to the level of the tide. The men were lucky enough to escape by the Iron Shaft. The Great Western Railway Company had so far been doing their own construction but the Great Spring was too much for them. They appointed John Hawkshaw as engineer and gave a single contract to Thomas Walker for the completion of the entire job. Thomas Walker's plan was to shut off the heading containing the spring by a curved shield braced across the Old Shaft, but there was 140 feet (43 m.) of water above this point. Extra pumps were brought in to lower the head of water to enable divers to work. The shield was in place on 24 January 1880 and the spring was isolated. Another shaft was dug beside the Old and Iron Shafts called the New Shaft. More pumps were put to work here and the water was lowered 121 feet (37 m.) in twenty-four hours, but then they could only hold it there because of water coming in from under the river.

Lambert, a diver, went down to shut a door and close a sluice in a bulkhead (a cross wall) 1,000 feet (305 m.) up the heading from the bottom of the Old Shaft, to stop the under-river water. He was working in a 30-foot (9 m.) head of water in pitch blackness and he needed three other divers to ease the movement of his air line. He set off with a crowbar and got to within 100 feet (30 m.) of the bulkhead when the weight of his air hose stopped him. Two days later he went down again in a Fleuss Dress which carried its own air supply in a cylinder. It was on his fourth attempt on 10 November that he succeeded in his mission to cut off the river heading from the Old Shaft, as he thought. But as the pumps did not achieve the expected drop in water, even more pumps were brought into play and the Iron Shaft was emptied. A foreman was thus able to get to Lambert's door to find that a valve on it had a left-hand thread so that the diver had opened it more instead of closing it. It was then closed, and the pumps could cope with the water from the Spring when the shield was removed. Men went up the Spring heading as far as the firm ground lasted, and built a brick bulkhead to seal off the Spring close to its source.

A few months later water again brought work to a halt. There was a pool of water, called Salmon Pool, left at low tide on the English side of the Shoots. In April 1881 it broke into the tunnel but did not fill all the works because the under-river headings had not yet met. At low tide Walker sent a line of men holding hands wading across the Salmon Pool and when one disappeared from sight he knew where the hole was. It was then sealed with a boatload of clay. Throughout the rest of 1881 and 1882 work proceeded smoothly. The pilot headings were broken out to full tunnel size and eventually lined with seventy-seven million bricks.

In 1883 the time came to tackle the Great Spring once and for all. Walker constructed a side heading to drain off the water while the main tunnel was being broken out and lined in this area, but it was not until the summer of 1885 that the lining was complete and the Great Spring was shut out.

As the Spring was being beaten the water had its final fling on a different front. On the night of 17 October 1883, the Severn Estuary experienced its highest tide since 1797, aided as then by a strong wind. A tidal wave swept over the Gloucester Marshes and poured down the 100-foot (30-m.) Marsh Shaft. Eighty-three men were in the works and they retreated up the long cul-de-sac of the under-river heading. The water rose to within 8 feet (2·4 m.) of the crown. The tide subsided, a dam was built round the Marsh Shaft, the water pumped out and the men rescued in a boat. The tunnel was finished in 1885 and on 5 September a Great Western train containing a select party ran through. Thomas Walker lost £100,000 on the contract and it had taken thirteen years to build, as long as the Mont Cenis Tunnel.

Thomas Walker discovered to his cost that traditional tunnelling methods were deficient in water-logged ground. New methods had to be devised, especially if the water-logged ground was soft and therefore fluid. The answer was a pneumatic tunnelling shield which is essentially an open-ended cylinder with a partition running across inside. Its major development work was done in the mud below the Hudson River in New York. The diameter of the cylinder is bigger than that of the final circular tunnel and it overlaps the lining of the tunnel so that the men working are always protected from the silt. The shield is jacked forward using the last ring of lining as a foot. Then the jacks are retracted, another ring of lining is put in and the shield is ready to go forward again. If it is being operated in water-logged ground, compressed air equal to the pressure of the water is pumped into the tunnel to keep the water out. It works on the same principle as a pneumatic caisson, except that a shield moves sideways instead of down-wards.

The map of the Hudson Tunnels shows the extensive system under New Jersey, New York and the Hudson River, as completed by Charles Jacobs in 1909. This was an extension of a scheme on which work had started thirty-five years previously for the uptown, or northern, pair of tunnels. It had been a long and difficult job because the Hudson river bed is filled with a near liquid silt which is up to 200 feet (60 m.) deep. The river itself is 40–50 feet (12–15 m.) deep.

When De Witt Clinton Haskin came to New York in 1874, rich from rail-way enterprises in the West, there was no crossing of the Hudson until

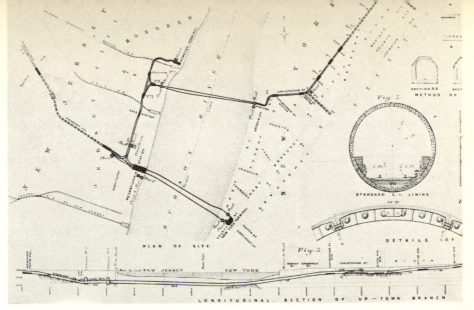

An extract from Charles Jacobs' paper delivered to the Institution of Civil Engineers after his completion of the Hudson Tunnels in 1909.

Albany, 175 miles (282 km.) upstream. He proposed to bore the (now) uptown pair of tunnels, using compressed air and canvas in the tunnel to support the silt without a shield. Anderson, his more experienced engineer, persuaded him to at least use timber to support the roof until the lining had been placed. A shaft was sunk in New Jersey and after some legal delay work started in compressed air in 1879.

A year later the river and silt broke through and twenty men were engulfed. A pneumatic caisson was sunk in the river, the bodies recovered and work resumed. This time the tunnel was shored by iron plates followed by a brick lining. A lack of public, railway and financial support brought

Cross-section of Pearson's tunnelling shield under the Hudson showing the slurry entering through the doors in the front face.

121

the work to a halt with over 2,000 feet (609 m.) of the north tunnel complete and 570 feet (174 m.) of its companion and parallel tunnel. It is amazing that Haskin's optimism had carried him so far.

The London contractors S. Pearson and Son took over in 1890 backed by English money and a pneumatic shield. Pearson's engineers found 250 feet (76 m.) of heading beyond a bulkhead full of slurry and a 12-foot (3·7-m.) diameter hole in the river. A large clay pudding contained in canvas was lowered into the hole from the river, the doors of the bulkhead opened and the slurry allowed to come into the tunnel dragging the pudding behind it to seal the hole in the river. The slurry was then cleared out of the tunnel and the shield erected. The tunnel was driven through the pudding and on, with the aid of the pneumatic shield. The air pressure was about 40 pounds per square inch (28,000 kg. per sq. m.) and men suffered many attacks of the "bends". Following the Baring Bank crisis of 1891 the company ran out of money and work had to stop in the following year. They sealed off the tunnel and filled the jacks on the shield with oil. They had added 1,800 feet (549 m.) of tunnel with a cast iron lining.

The East River Gas Company were building a 10-foot (3-m.) diameter tunnel under the East River at this time. It was 125 feet (38 m.) below the water level and they therefore assumed that it would go through rock. It did not and the contractor gave up. Charles Jacobs was called in and he installed pneumatic tunnelling shields to complete the tunnel in 1894. Jacobs was thus the expert at tunnelling in New York water-bearing material. In 1895 he examined the Haskin-Pearson tunnels. He found them to be full of water to within 11 feet (3 m.) of the top of the New Jersey shaft. This water was pumped out and he reported that 3,916 feet (1,194 m.) of the north tunnel from New Jersey and 160 feet (49 m.) from New York, were in reasonable condition, and the same for the 570 feet (174 m.) of the south tunnel. The shield left by Pearson in the north tunnel was also in good condition: indeed, this shield went across the river to complete the north tunnel in 1904 after work was resumed in 1902 by Jacobs.

The traffic movements were examined between New York and New Jersey and it was decided to extend the (uptown) tunnels to run from a station at Hoboken, and right on up Sixth Avenue via Morton Street and Christopher Street. A pair of downtown tunnels would also be needed running from a station in Church Street New York, to under the Pennsylvania Railroad's station in New Jersey linking with the uptown tunnels, about 12 miles (19 km.) of tunnel in all. Jacobs pushed the Pearson shield on 136 feet (41 m.) until a rock ledge was reached and the difficult job of driving a tunnel half in rock and half in silt was started in November 1902. This was achieved by erecting an apron ahead of the shield to protect the men placing the charges in the rock. There was a blow (an escape of compressed air from the tunnel to the river) and clay was dumped from the river to consolidate the silt. This failed and Jacobs hit on the idea of hardening the clay by using kerosene blowlamps on it. "The first time men ever made bricks at the bottom of a river," he said.

An astounding incident concerning a "blow" occurred in 1905 on the Brooklyn–Manhattan Transit Company's 6,790-foot (2,070-m.) tunnel under the East River. Richard Creegan was on guard looking out for leaks at the tunnel face. He saw one starting and stuffed a hay bale, kept for the purpose, into it. It was swallowed up immediately. He put in a second bale but his hand got caught and he was sucked in. His kicking legs were visible for a few seconds in the roof of the tunnel as his mates tried to pull him back. Then he was blown by the compressed air right through into the river, where he was rescued alive.

Jacobs' brick-making operation was successful and the shield went on a further 740 feet (226 m.) to meet the bulkhead, left by Pearson in the tunnel from New York, on 11 March 1904. The new tunnel is 16 feet 7 inches (5 m.) in diameter, which is 3 feet (91 cm.) less than the old tunnel. New shields were erected for the south uptown tunnel and the downtown pair of tunnels. They were driven by sixteen 8-inch (20-cm.) hydraulic jacks using water at a pressure of 5,000 pounds per square inch (352 kg. per square m.). Often the shields went ahead without admitting any silt at all, but usually about five per cent of the displaced material was admitted through doors in the shield face because the shields could be better steered thus. On one occasion the uptown south tunnel was driving blind when it slowed, because it had entered stiffer material, so the superintendent opened one of its doors to admit this. For half an hour progress speeded up, when suddenly a column of liquid silt shot into the tunnel engulfing one man. The rest ran for their lives to behind the safety of an emergency door in a bulkhead. A weighted canvas sheet was lowered into the river and drawn on to the shield by allowing silt into the tunnel by the emergency door. This silt was removed for eight days until the canvas was sealing the shield and it took a further nine days to recover the heading. The south uptown tunnel was finished in August 1905 and the downtown pair of tunnels were finished four years later.

The Morton Street to Sixth Avenue tunnel was constructed with a shield and it went on up Sixth Avenue by cut-and-cover methods, but great care had to be taken to avoid undermining nearby buildings only 15 feet (5 m.) away, and the elevated railway. The uptown and downtown link tunnels were also done with a shield except for the junctions in the uptown triangle. 123

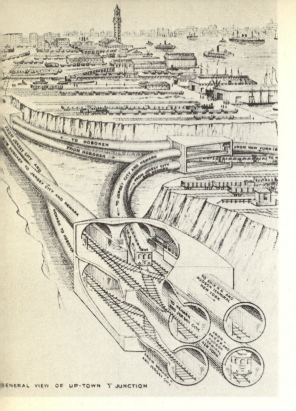

GENERAL VIEW OF UP-TOWN Y JUNCTION

A drawing of the Uptown Junction, New York, taken from Charles Jacobs' paper. The double-deck junction boxes were built on the surface and sunk as pneumatic caissons.

These double-decked junctions were built on the surface and sunk as pneumatic caissons. They took three months to sink to their levels of around 90 feet (27 m.), and in one the wreckage of two long-ago-sunken canal boats had to be cut up and passed through the air locks with the rest of the displaced material.

Of the 40,000 men who passed through Jacobs' air locks, 1,573 suffered from compressed air sickness, or the "bends". However, only three died because of the pioneer work done on men working in compressed air, that had been carried out during the construction of Eads' St Louis Bridge in 1870.

The rush of spectacular railway engineering was obviously in the nineteenth century because that was when railways were first built, but railway engineers are still at work, more quietly now and with less publicity. The New Woodhead Tunnel shows that the unexpected still happens and an engineer needs more than his scientific training.

When it was planned to electrify the line between Manchester and Sheffield, it was obvious that the old Woodhead Tunnels (page 90) would have to be altered to accommodate the overhead electric cables. The tunnels were also deteriorating and maintenance was difficult with more than eighty trains a day in each direction, so it was decided that the cheapest solution

Right: *Sinking a shaft on the moors above Woodhead for the new tunnel in 1950.*

Inside the pilot tunnel for the new Woodhead Tunnel near the Woodhead end in 1951.

would be to build a new twin-track tunnel. The consulting engineers for the New Woodhead Tunnel were William Halcrow and Partners and things were ready for the contractor, Balfour Beatty and Company, to start work in February 1949, with John Isdale Campbell in charge. At 3 miles and 66 yards (2·89 km.), 44 yards (40 m.) longer than its predecessors, the New Woodhead Tunnel is the third longest in Britain (the Severn is the longest) and straight except for a curve at the Woodhead end to make joining the line to Manchester easier.

A 16-foot (4·8-m.) diameter shaft was put down 467 feet (142 m.) from the moors above to the tunnel level and, working from four faces, the portals and the shaft bottom, the 12-foot (3·7-m.) square pilot tunnel was through by 16 May 1951. Only one shaft had been used as opposed to the five used on the original tunnels, because it was planned to enlarge the pilot to full size by radial drilling instead of the normal face drilling. Radial drilling gives a much bigger working area and not so many shafts are needed to give a good rate of construction. The pity was that radial drilling did not give a pre-dictable tunnel profile in this rotten rock and the idea had to be reluctantly

Roof fall in the Woodhead Tunnel. After the debris had been cleared there was a 100-foot (30-m.) high cavity above the line of the tunnel.

abandoned. Therefore several intermediate enlargement chambers had to be opened up in the pilot tunnel to maintain the desired rate of progress, and there was some chaos where the spoil from one had to pass through another. The problem was delayed and the solution suggested by a roof fall, 900 feet (274 m.) from the Woodhead end. It was observed on 4 June 1951 that the ribs supporting the enlarged tunnel were bending; they were strengthened. The ribs buckled again after four days and 76 feet (23 m.) of roof fell in. It took six months to clear the debris of the fall and, so that this would not hold up the work in other parts, a by-pass tunnel was built. It was found that this by-pass tunnel gave two enlargement faces without the spoil from one interfering with work on the other. The answer to the spoil cross-over problem was thus discovered and eventually 9,500 feet (2,896 m.) of by-pass tunnel were built to give nine enlargement working faces.

In driving the pilot tunnel thirty-two holes were needed in its face for the gelignite charges. The full face required fifty-five holes between 6 and 10 feet (1·8–3 m.) deep. Half a million cubic yards of rock were removed, each needing $2\frac{1}{2}$ pounds (1 kg.) of gelignite on average. Usually about 100 gallons (455 litres) of water per minute had to be pumped out of the works but it rose to 1,000 gallons (4,546 litres) per minute at one stage when the pilot tunnel entered some saturated sandstone. The pumps coped with this

A by-pass tunnel used during construction of the new Woodhead Tunnel.

Pilot tunnel and full size enlargement in the Woodhead Tunnel. The pilot was 12 feet (3.65 m.) wide and the enlargement 30 feet (9 m.). Blasting holes can be seen in the enlargement face.

easily, but the water caused some trouble when the lining was being concreted. The moulds for the concrete lining, called shuttering, were mounted on a carriage, 100 feet (30 m.) long at the Woodhead end and 70 feet (21 m.) long at the Dunford Bridge end. The original plan for the lining was to work from each portal following 500 feet (152 m.) behind the radially drilled enlargement. The radial drilling had been abandoned, but using these shutters the lining could still only proceed in this fashion and it had to be speeded up. After initial practice, the shutters running on two narrow-gauge tracks, could be set up, concreted and moved five times in two weeks. The overbreak (extra material removed over and above the planned tunnel profile) was on average $17\frac{1}{2}$ inches (45 cm.) which resulted in fifteen per cent more spoil to be removed and an extra eighty-five per cent of concrete for the lining, which is at least 21 inches (53 cm.) thick, and 5 feet (1·5 m.) thick through the 100-foot (30-m.) high cavity left by the roof fall. A minor roof fall on 21 May 1953 delayed the final movement of the shutters for two weeks. The track was laid by the end of that year. New Woodhead Tunnel had cost four and a half million pounds and the lives of six men.

Tunnelling is the most dangerous pursuit of the civil engineer, because there is no escape if anything goes wrong. A man falling off a bridge under construction stands a chance of survival if he falls into the river, but in a tunnel he is buried alive. Adding to the danger is the fact that the tunneller does not know exactly the nature of the ground he is about to enter: the predicted clay can change to water-logged gravel, an unknown spring can be tapped, apparently sound rock can suddenly collapse owing to unseen and unsuspected flaws. A tunnel engineer is to some extent working in the dark.

10

Modern Road Works

A complete rethinking had to take place on road building with the advent of the motor car and pneumatic tyres. If the roads of the United States and Europe in the nineteenth and early twentieth centuries were properly metalled at all, they were macadamized. The 8-inch (20-cm.) layer of broken stones were ground and consolidated by the action of metal-rimmed wheels and water. Water carts and horse-drawn rollers were used after 1830 to speed up this process. The stone crusher and steam roller came into widespread use after 1860 and greatly facilitated the building of macadamized roads. The roads were maintained simply by putting on more stone and rolling it. But the pneumatic tyres of a motor vehicle sucked the roads instead of compacting them, thus quickly destroying the metalling and causing great clouds of dust. House values were halved on some main roads and the British medical journal *The Lancet*, spoke of "the dust menace" as a serious health problem. The dust was not good for motors or motorists either. A Royal Commission was held in 1906 on dust prevention and the RAC formed a dust committee. In 1907 the Roads Improvements Association organized a dust laying competition on a road near Reading. In 1908 the French Minister of Public Works called an International Congress in Paris. The conclusions of the various deliberations were: that tar should be applied to existing macadamized roads, that new roads should be made of concrete or asphalt, and that central government must take charge of all main roads. None of these ideas was new, but the need for them was. By 1908, 1,269 miles (2,042 km.) of English roads had been treated with tar, hand-painted at first, but soon a technique of spraying with compressed air was developed. It was discovered in France that hot tar penetrated further, binding the whole top layer of stones and lasting longer.

The same investigations were being held in the United States. A census in 1904 revealed that of the 2,151,570 miles (3,464,000 km.) of rural highways (i.e. not city streets), only 152,662 miles (245,000 km.) had some kind of metalling, the best of which was 38,622 miles (63,000 km.) of water-bound macadam. This compared with 220,112 miles (354,000 km.) of railway then. By 1914 10,500 miles (17,000 km.) of the macadamized surfaces had been sprayed with oil or tar, but such bituminous roads, although dust proof, cannot take heavy loads. Concrete or asphalt was needed and Wayne County, Michigan were the pioneers of building roads for loads: between 1909 and 1914 they built 1,500 miles (2,414 km.) of the 2,348 miles (3,779 km.) of concrete roads in the entire country. After World War I, the value of motor transport was properly appreciated and by 1924 there were 31,188 miles (50,000 km.) of concrete roads, construction proceeding at 6,000 miles (9,600 km.) a year. In the same year, there were also 9,700 (15,600 km.) miles of asphalt, showing the American preference for concrete, 45,000 miles (72,400 km.) of bituminous macadam, 60,000 miles (96,500 km.) of water-bound macadam, 310,000 miles (500,000 km.) of gravel, and over $2\frac{1}{2}$ million miles (4 million km.) with no metalling at all. This large proportion of unmetalled roads should not be exaggerated. Even in 1947 forty-eight per cent of United States' roads were not made up; this is because eleven per cent of the roads carry seventy-four per cent of the traffic.

The early 1920s saw a new approach to highway design: roads were designed specifically for motor traffic and higher speeds, with long sight lines, gradual curves, banked corners, dual-carriageways and no intersections with other roads. The motorway was born. Some of the ideas came from the earlier city streets where heavy congestion had demanded a solution. North-eastern Boulevard, now Roosevelt Boulevard, in Philadelphia was started as an 8-mile (13 km.) section in 1903. Its 300-foot (91-m.) width was occupied by a central 60-foot (18-m.) wide through road, two 34-foot (10-m.) wide side service roads, footpaths and lawns. Boston, too, had such boulevards, and after 1910, Grand Boulevard in the Bronx, New York, had some of its crossing roads going beneath street level. In 1906 Eugene Henard published a tract on "Intersections having superimposed highways" giving examples from Paris and other European cities. In it he proposed the "cloverleaf" junction and the traffic island. The state of New Jersey took the lead in superhighways between towns in the early 1920s and used the first road island on the eastern approach to the Camden–Philadelphia Bridge. The Camden project also contained the first two-level junction on a state highway. The first cloverleaf junction was opened in 1928 at Woodbridge on the New Jersey Turnpike. New Jersey also had more miles of dual-carriageways than any other state at this time.

The Germans and the Italians were not far behind. A 53-mile (85-km.) concrete road from Milan to the Italian lakes was opened in 1925 after two

Autobahn between Frankfurt and Heidelberg in the late 1930s. The motor car caused a complete rethinking on road building.

years of work. The first autobahn could be an urban road known as the "Avus" in Berlin in 1921, although the 1929 Cologne–Bonn Autobahn was the first to link two towns. There were no junctions on this 12-mile (19-km.) concrete dual-carriageway road. When the National Socialist Party came to power in 1933 they announced a programme for the building of 4,300 miles (6,920 km.) of autobahn, and Hitler cut the first sod on the Frankfurt–Mannheim–Heidelberg road. By 1937, 950 miles (1,529 km.) were finished, 1,120 (1,802) were under construction and 1,150 (1,850) were ready for construction to start. They were on brand new routes, by-passing towns and designed for 100-mph (160-kph.) travel. The four-lane dual-carriageways were 80 feet (24 m.) wide.

In 1930 a new road was started through the Grossglockner Pass in Austria, from Bruck in the north 50 miles (80 km.) south to Dollach. The whole route is over 6,000 feet (1,829 m.) above sea level rising to 8,000 feet (2,438 m.). This was a time of severe unemployment and the labourers were hairdressers, clerks, etc., who found working in such conditions harsh. One in fact died of heart failure climbing through the snow before he even got to the site. Work was only possible for five months of the year at the most.

Frank Wallack planned and built the road, and when he was out surveying he found traces of an old Roman road following nearly the same line, levels and gradients. Engineers two thousand years apart were coming to the same conclusions. The area passed through is the Höhe Tauern, where a thousand years previously gold and other metals had been mined. Near the entrance to an old mine there was a great heap of rocks in the path of the road. These were dug out to reveal 4 feet (1·2 m.) of black ice. When

133

that was removed there were more rocks, then ice and more rocks again. These were levelled for the road, but in two weeks the path of the road became mud and slush. The rocks were dug out to reveal more black ice melting below. Drilling down Wallack found alternate layers of ice and rocks. The rocks were the summer excavations of the medieval miners and the ice was the winter snow. Each layer represented a season of work. The whole lot had to be dug out and filled in before the road could proceed. The bones of a miner were found in one of the rock layers and they moved him to a cemetery in Heiligenblut. Elsewhere work was hindered by avalanches and landslides. In one place a warm wind from the south melted a 7-foot (2 m.) layer of snow in a day and the whole hillside came down, trees, farm buildings and all. The road was eventually finished in August 1935. Even now it is only open for about 100 to 160 days a year.

At the same time in England, a tunnel under the Mersey Estuary was being built. Its diameter was to be the biggest in the world. Liverpool, Birkenhead, Wallasey and Bootle form an economic unit which had been growing during the Industrial Revolution. By the end of the nineteenth century the River Mersey, which had brought their prosperity, was becoming a hindrance to communication between the towns. A railway tunnel was built in 1883, and this and the ferries had to suffice until after World War I, when the increasing numbers of motor cars demanded a better crossing. A

A junction in the Mersey Tunnel under construction.

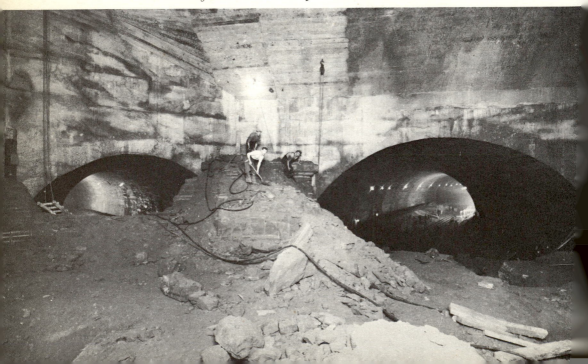

The junction between the main and Birkenhead dock branch tunnels.

committee was formed in 1922 and engineers Basil Mott and Maurice Fitzmaurice reported that a tunnel was the best answer. They proposed a 44-foot (13·4-m.) inside diameter tube to take four lanes of motor traffic with two tram tracks underneath. (Space for the latter was provided when the main tunnel under the river was built, but never used.) There were to be two entrances on each side of the river.

A contract was placed in December 1925 and work started immediately by sinking two shafts at each end of the under-river portion of the main tunnel. They were 21 feet (6·4 m.) in diameter and went down for 190 feet (58 m.) through sandstone. Sandstone is porous but that on Merseyside was also cracked and fissured, so the water flow had to be checked. This was done by boring 2-inch (5-cm.) holes ahead of the work and pumping in sodium silicate and ammonium sulphate which formed a gel and sealed the pores of the rock. Then liquid cement was pumped in to fill the cracks and fissures. This cut the water flow by ninety per cent. Two pilot tunnels were started riverwards from half-way down each shaft to meet under the river on 3 April 1928 with an error of only 1½ inches (3·8 cm.). Each were 12 feet (3·7 m.) high by 15 feet (4·6 m.) wide. The lower headings were kept 150 feet (46 m.) in advance of the upper headings and the ground ahead was tested by boreholes up to 100 feet (30 m.) long in front of and above the lower headings. Water drained down from the upper to lower headings. Another   135

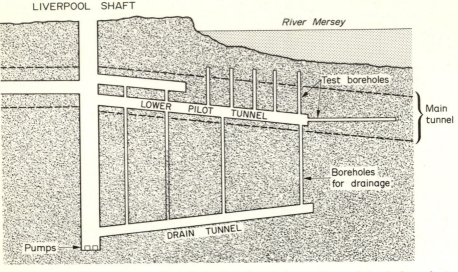

LIVERPOOL SHAFT

River Mersey

Test boreholes

LOWER PILOT TUNNEL

Main tunnel

Boreholes for drainage

DRAIN TUNNEL

Pumps

Cross-section of Mersey Tunnel showing the relative positions of the shafts and pilot tunnels.

tunnel, 7 feet (2·1 m.) in diameter, was driven from the bottom of the Liverpool shaft to meet the pilot tunnels under the river, and water drained from the lower pilot tunnel into this, and thence to the bottom of the shaft from which it was pumped. There was not much water on the Birkenhead side so the drainage tunnel was not needed there.

Now the pilots had to be opened out to make the full-size tunnel of 46 feet (14 m.) diameter. The top half was done first and lined, the excavated material being dropped down chutes into the lower pilot tunnel where it was carted off on an electric railway. Then the bottom half was opened out and the rock taken away by a railway suspended from the roof. The enlarging was carried out at several points simultaneously. Thus it was very convenient to have the railway tracks out of the way of the work. The cast iron segments of the lining were placed by a special hydraulic machine. Big gaps behind the lining were filled with rocks and then liquid cement was pumped in to fill the smaller voids uniting the lining and the rock.

That was the circular under-river section of the main tunnel finished. It is just over a mile (1·6 km.) long through sandstone all the way. The average rock cover is 20 feet (6 m.) and the lowest part is 170 feet (52 m.) below the high water mark. At the same time work had been progressing on the landward parts of the main tunnel, and on the two-lane dock branch tunnels of 26 feet (8 m.) diameter on each side of the river. These were made

a bit more than semi-circular with air ducts beneath because by now the

Work proceeding on the Mersey Tunnel. The upper half has been lined and the lower pilot tunnel is being opened out to full size and lined. The spoil is taken away by the railway suspended from the roof.

Erection of the reinforced concrete roadway in the main Mersey Tunnel.

plan to run trams under the road had been abandoned. They were built by opening out a pilot tunnel, or with a tunnel shield, or by cut-and-cover methods, depending upon the situation and the type of ground. All the tunnel walls were sprayed with a concrete composition and bitumen emulsion, making them virtually watertight.

The reinforced concrete roadway was installed supported on walls in the circular tunnel and on pillars elsewhere. The road in the main tunnel is 36 feet (10·9 m.) wide, 19 feet (5·8 m.) in the dock branches, with 16 feet (4·9 m.) headroom. The original surface was paved with cast iron blocks but now it is asphalt. The ventilation problem was unprecedented. A 1,000-foot (304-m.) completed section was bricked in, and damp hay encouraged by petrol was burned to see how quickly the smoke could be extracted. These experiments and other calculations led to the planning of six ventilating stations containing fans of up to 28 feet (8·5 m.) diameter to put in and extract 2,250 tons (2,286,000 kg.) of air per hour. All was finished after nine years for George V to open the tunnel in July 1934. He named it "Queensway". One million two hundred thousand tons (1 million, million kg.) of rock, clay and gravel had been removed with the aid of 560,000 pounds (250,000 kg.) of gelignite. The total length of the tunnels, including the dock branches of 3,215 feet (980 m.), is 2¾ miles (4·4 km.).

A second Mersey Tunnel was completed in February 1974. It is almost 1½ miles (2·4 km.) long and consists of two tubes, each 32 feet (9·7 m.) inside diameter for two lanes of traffic. A 12-foot by 12-foot (3·7 m.) pilot tunnel was completed to explore the ground at the end of 1966. This was opened out to nearly 34 feet (10·4 m.) diameter by a mechanical mole for one of the tubes. The mole was 48 feet (14·6 m.) long with a disc of cutters at the front which revolved once a minute. The machine was powered by ten 100-horsepower electric motors. It also placed the concrete segments of the tunnel lining and passed the excavated rock to its rear. This first tube was used by two-way traffic while and until the second was built, also using the mole.

The same problem of physical communication that existed between Liverpool and Birkenhead has occurred in many places throughout the world. Istanbul and Uskudar in Turkey are now linked by the Bosporus Suspension Bridge with a span of 3,523 feet (1,074 m.). The bridge was opened in 1973 and it is amazing that it was not built much earlier. San Francisco and Oakland were separated by San Francisco Bay, and while men were tunnelling under the Mersey, the Bay Bridge was under construction.

The Bay Bridge is actually two suspension bridges joined end to end at a common anchorage point, a 540-foot (165-m.) tunnel through Yerba Buena Island, and a 1,400-foot (426-m.) span cantilever bridge to the mainland. The whole adds up to a 4¼-mile (6·8-km.) crossing on two decks, six lanes of traffic on the upper deck, and three lanes for heavy traffic and two urban train tracks on the lower deck. The suspension bridges have a span of 2,310 feet (704 m.) each. The common anchorage block between the two suspension bridges is interesting because it is the second deepest concrete monolith in the world. It goes down 228 feet (69 m.) on to a sloping rock surface. The rock was levelled by dropping a pointed heavy steel weight down the fifty-five wells in the monolith through which the displaced material was taken out by a grab. As the monolith sank, it was built up on top to keep it above the water. When it was down the wells were filled with concrete. (The deepest monolith is on the Hawkesbury Road Bridge in Australia which goes down to 233 feet (71 m.). At 129 feet (39 m.) it suddenly plunged 53 feet (16 m.) and disappeared from sight.) The Chief Engineer for the Bay Bridge was C. H. Purcell and it was opened in November 1936 after three years of construction.

One of the most audacious highways in the world is the Caracas Autopista

The Caracas Autopista. The old road can be seen in the background.

in Venezuela and it is fitting to close this chapter with some mention of it, because it shows the confidence that road engineers have reached. The Caracas Autopista strides almost contemptuously and directly across some very difficult terrain. Caracas is the capital of Venezuela and its only link with the port of La Guaira, 10 miles (16 km.) away on the Caribbean Sea coast, is by road. The old road was $18\frac{1}{2}$ miles (29 km.) long and overloaded with six thousand vehicles per day. Traffic jams built up and the journey could take many hours, although only one hour on a clear road. Ninety per cent of the road comprised 395 bends with radii down to 50 feet (15 m.). If this was not enough, there was the final hazard of landslides which sometimes removed all or part of the road, or at other times blocked it.

Engineers started work on the new road in 1950, after six years of planning and surveying. Opened in 1955, it is $10\frac{3}{4}$ miles (17·3 km.) long and the drive now takes fifteen minutes because the two 12-foot (3·7-m.) dual-carriageways, designed for 50-mph (80 kph) speeds, have only thirty-six bends on a minimum radius of a 1,000 feet (300 m.). There are two tunnels, one a quarter of a mile (400 m.) long and the other just over a mile (1·8 km.). The road is in cuttings, on embankments or viaducts, or on steps blasted along the sides of deep ravines for virtually its full length.

One of Freyssinet's bridges on the Caracas Autopista under construction. The middle part of the wooden centring is being raised from the valley floor.

There are also three major bridges, all reinforced concrete arches and with spans of 499, 479 and 452 feet (152,146 and 138 m.) respectively. The 70-foot (21-m.) wide road is supported on spandrel columns above the arch in each case. The arches are made of three hollow rectangular section ribs with a rise of 108 feet (33 m.). Eugène Freyssinet, the designer of the bridges, could not use the normal practice of supporting the wooden formwork, or centring, for the arches from the valley floor because 100-mph (160 kph) winds blow down those gorges. In addition, the road was to be over 200 feet (61 m.) up and the sides on the valleys were very crumbly. The formwork for a quarter of the spans each side was erected and tied back in position to the abutments. A thin layer of concrete, the bottom of the arch-to-be, was laid over these to give the extra strength required to take the weight of the centre half-span of formwork which was raised entire from the valley floor where it had been built. The French contractors, Campenon-Bernard, took two days over this operation. When it was in place on the two quarter spans, a layer of concrete was laid overall and allowed to set. Then another layer, each layer acting to some extent as centring for the layer above.

11

Dams and Hydro-electricity

Water is the most powerful of the elements that civil engineers have to deal with, therefore dams are the biggest things that they have to build. They have been building them for a long time: the civilization of ancient Egypt would not have come into existence without them. They were of great agricultural and strategic value, as has been observed in Chapter 1. All the early dams were made of earth, then as they got bigger and higher in the late nineteenth century, masonry was preferred, as in the Aswan Dam. Today, with the development of soil mechanics as a science, there is a return to the earth dam; the new Aswan High Dam is of this construction.

Basically, there are two types of dam: gravity and arch. All earth dams are gravity dams; they depend upon their weight only to hold them down, and to resist the tremendous push that a head of water can give. The arch dam is curved upstream and resists the thrust of the water in the same way that an arch bridge takes a load, but sideways. It is therefore most suitable in moderately narrow rock-sided gorges. Most arch dams also have an element of the gravity type in them. The Boulder Dam is a good example. Both types of dam must have a good impervious foundation, preferably right into solid rock, because it is at the base of a dam that the water pressure is greatest, and once water starts to flow, it erodes a bigger and bigger passage and there is no stopping it until the reservoir is empty. The Puentes Dam had this trouble.

The Puentes Dam was built during the years 1785–94 across the River Guadalentin in Spain. It was 164 feet (50 m.) high and 925 feet (282 m.) long. It had been intended to found the dam on solid rock, but there was a deep pocket of earth in the centre of the valley and piles were used in this part of the foundations. For eleven years the water was never more than 82 feet (25 m.) deep. Then on 30 April 1802 it rose to 154 feet (47 m.). At 2.30 p.m. it was noticed that the water downstream was very red, the colour of the sub-soil beneath the dam. At 3.00 p.m. there was a muffled explosion from within the dam, then an enormous mass of piles and timbers forced their way out from under the dam, followed by a mountain of water. Within

142

the hour the reservoir was empty and 608 people in the valley below were dead. The top of the dam remained intact over a gaping arch, 56 feet (17 m.) wide and 108 feet (33 m.) high. The saturated sub-soil around the pile foundations could not take the pressure due to a 154-foot (47-m.) head of water, which was over 4 tons per square foot (45,000 kg. per square m.).

It was the Spanish, in fact, who were the first to develop the large-scale masonry dam. Their earliest, and still in use, is the Alicante Dam built from 1579–94. It is 135 feet (41 m.) high and closes the Gorge of Tibi. These dams were for irrigation, but in the late nineteenth century, as industrial towns grew bigger, dams were built to impound water for town supply.

The 184-foot (56-m.) high Furens Dam for St Etienne in France was completed in 1866, and the first in Britain was the Vyrnwy Dam in North Wales to provide water for Liverpool, occupying the years 1882–90 in its construction. New York got its first water in 1842 from a dam across the Croton River. Old Croton Dam was 50 feet (15 m.) high and in 1905 was submerged in the reservoir behind the 297-foot (90-m.) high New Croton Dam. But New York was still thirsty and in 1908 started on a scheme to bring water from the Catskill Mountains, over 80 miles (129 km.) away to the north of the storage reservoir at Kensico, which itself was 32 miles (51 km.) north of Long Island. The dam at Kensico has a maximum height of 307 feet, (94 m.), and is 235 feet (72 m.) thick at its base tapering to 28 feet (8·5 m.) at the top. Stone was quarried about a mile (1·6 km.) away to the east using very large blasts. One notable blast occurred on 20 March 1914. An area of 100 feet (30 m.) by 900 feet (274 m.) was blown out from 142 holes with an average depth of 40 feet (12 m.). Thirty tons (30,000 kg.) of dynamite were used and the heaviest charge in one hole was 1,400 pounds (635 kg.). This blast broke up about 120,000 cubic yards of rock into pieces small enough to be handled by steam shovels.

All these dams were gravity dams, except the Furens which to a small extent used the arch principle as well, and were built of cyclopean concrete, also called rubble masonry, inside a facing of cut stones or concrete blocks. Cyclopean concrete is so called because it is like concrete mixed by a giant. Some of the stones in it are so big that they have to be laid individually like crude masonry, using ordinary concrete as a mortar. In the Vyrnwy Dam, for example, some of the stones in the rubble interior weighed 10 tons (10,000 kg.) and half of them were over 2 tons (2,000 kg.). The big stones were laid by seven steam cranes and bedded into the mortar with wooden mallets. Blunt swords were used to ram smaller stones and mortar into the interstices. In this fashion the dam was brought up to its maximum height of 161 feet (49 m.) above its foundations and used 510,000 tons (500 million kg.) of masonry. The village of Llanwddyn was drowned by the reservoir and a new village was built below the dam. Even the bodies were moved from the graveyards. Masonry was preferred for these dams because the techniques of soil mechanics were not yet sufficiently advanced to permit such big dams to be built of earth, and perhaps also because public faith had been eroded by a series of disasters with earth dams in the last half of the nineteenth century. The worst of them was the Johnstown disaster of 1889.

Johnstown had a population of 30,000 in 1889; it was a steel works town on the Pennsylvania State Canal. The canal had always been short of water in the summer months, so in 1852 a supply reservoir had been formed by damming the Little Conemaugh River at South Fork, 15 miles (24 km.) up the valley from Johnstown, and 450 feet (137 m.) above the town. The dam had taken fifteen years to finish, because money was short but it was properly made. It consisted of an earth and rock fill bank, 72 feet (22 m.) high and 930 feet (283 m.) long. It was 20 feet (6 m.) thick at the top and

Wreckage in Johnstown after the South Fork Dam had burst in 1889.

270 feet (82 m.) thick at its base. There was a 72-foot (22-m.) wide spillway cut out of the rock of the hillside at one end and 11 feet (3·3 m.) below the top of the dam. There were five 2-foot (60-cm.) diameter cast iron pipes under the base of the dam to drain the lake if need be. When full, the lake was about 5 miles (8 km.) round and held over twenty million tons of water. The Pennsylvania Railway came to Johnstown six months after the dam was completed and rendered the canal obsolete. The dam was forgotten, but it was eventually bought by the South Fork Fishing and Hunting Club in 1879. Without engineering advice, they repaired it but left a sag of 3 or 4 feet (1 m.) in the middle and most vulnerable part of any dam, they lowered it by 2 or 3 feet (1 m.) to make the road over it wider, and worst of all, they sold the cast iron pipes and valves under the dam for their scrap value. All this meant that if more than about 4 feet (1 m.) of water went over the spillway, the dam would overflow and erode at the centre, and there was no means of lowering the water level now the cast iron pipes had gone. A disaster was inevitable but no one wanted to know. On Friday, 31 May 1889 at 11.30 a.m., after days of torrential rain, the danger level was exceeded and the dam overflowed. It burst at 3.10 p.m. and an hour later a 36-foot (11-m.) wave of water hit Johnstown. Three thousand buildings of brick and wood moved with the torrent and piled up in a compact mass against a bridge across the valley. Then in the evening the whole mass caught fire. In this disaster 2,209 people were killed.

However, faith in earth dams has now been restored and what is notably the most important dam in the world is an earth dam. A whole nation depends on the Aswan High Dam for its existence. In 1970, it replaced the Aswan Dam, which had played the same role since 1902. Before 1902 Egypt had always used basin irrigation, that is, when the Nile flooded in July the water was kept lying on the land in basins by small dykes. A single crop was then planted in the sodden soil after the flood had gone down. The Aswan Dam enabled the Nile valley to go over to perennial irrigation, that is, water could be let on to the land in canals from storage reservoirs. This gave two or three crops a year. The dam was fitted with 180 sluices and these passed the main part of the silt-laden flood water into the old basins. At the tail end of the flood in late November, when the water was clear, the sluices were closed and the reservoir filled up during the months of December, January and February. Water could then be drawn off into the canals when needed.

William Willcocks, an engineer in the Egyptian Department of Irrigation, surveyed the Nile and submitted his plans for an Aswan Dam to an international commission which included Benjamin Baker, of Forth Railway

Bridge fame. The site at Aswan was ideal for the dam because here the river left the mountains of southern Egypt and entered the flatter lands 600 miles (966 km.) from the Mediterranean. The river flowed in five channels over a hard granite ridge. Willcocks' original plan was for five separate dams joining the islands, but Baker changed it to a single straight wall 1¼ miles (2 km.) long holding a maximum 67-foot (20-m. depth of water. In February 1898, John Aird contracted to build the dam within five years. Preliminary work occupied most of the first year. A town was built for 15,000 workers and railways laid to the quarries. All the materials had to be imported, except for the granite which was the same stone and came from the same quarries as the Pharaohs had used.

His Royal Highness the Duke of Connaught laid the foundation stone early in 1899, and in the same year work began on the sudds. These were the coffer dams built above and below the site of the dam proper. The stone sudds downstream were built first so that the sand sudds upstream could be made in the resultant still water. The stone sudds were closed one by one from the eastern bank. As each of the five channels got narrower, the water rushed faster through the decreasing gap until in one, 4-ton (4,000-kg.) stones dropped in by crane were carried away. Aird solved this problem by filling pairs of railways trucks with 50 tons (50,000 kg.) of stone and running them into the gap to be closed. The stone sudds were 30 feet (9 m.) wide at the top and 120 feet (36·6 m.) wide at their base. Three channels were closed by the end of 1899 and as soon as the flood had gone down below the stone sudds, the sand sudds were easily built. The granite sand was in bags. The sand sudds were 10 feet (3 m.) higher than the stone sudds, at 60 feet (18 m.). The water was then pumped out from between the sudds and work on the foundations started. This entailed much more work than was expected because some of the granite was rotten and the engineers had to dig deeper than anticipated, 40 feet (12 m.) deeper in places, to find rock on which to stand the dam, 100 feet (30 m.) wide at its base. There was a low flood that year and work could continue earlier than expected. Aird took the gamble of building in four channels, diverting all the Nile into the one remaining passage. More rotten granite was found and work went on at night as well under electric arc lamps with only three months to go before the next flood, by which time the rubble masonry wall had to be high enough to withstand the water.

The *Daily Mail* described the scene: "Here and there, under broad sun-helmets, like tall mushrooms, may be found a wily Greek or excitable Italian, acting as a useful lieutenant to, and directing the work being executed by, the solitary Englishman perched yonder on an elevation of

Water flowing over the partially-constructed Aswan Dam in August 1900.

masonry, apparently an idle spectator, and yet seeing all, and occasionally acting as judge in the many disputes arising between the different factions." August was a tense month as the water flowed steadily over the wall. The engineers could only wait. When the water subsided, all was well. The flood had been the lowest ever recorded. The fifth channel was closed by February 1901 and the next flood was greeted by a complete wall from bank to bank. The granite top was finished, the sluices installed and the navigation locks cut in the western bank by June 1902, a year ahead of schedule. The height of the dam was over 100 feet (30 m.) from the bottom of its foundations, and its width over the main section containing the sluices was 23 feet (7 m.). When full, the reservoir went back for 143 miles (230 km.). The dam was opened by a silver key and much junketing on 12 December 1902. Egypt was back on her feet and the country began to prosper.

Shortly after the dam was in use, it was found necessary to build an apron in front to stop the water from the sluices scouring the river bed and undermining the foundations. All the loose rock was cut out, the hollows filled up with masonry and built up to the sluice level. It went for 200 feet (61 m.) downstream and was 37 feet (11 m.) thick in places. It was finished in 1906 and contained 350,000 tons of masonry, compared with one million tons in the dam and five million in the Great Pyramid.

Soon the Egyptian Government regretted that the dam was not higher, and in 1907 Baker and Aird were commissioned to enlarge it. The engineering difficulty was the expansion and contraction of one and a quarter miles (2 km.) of stonework as the temperature changed from 150°F (65°C) in the day to 40°F (4°C) on the coldest nights. The old dam with five years of maturity would not behave in the same way as the new masonry. Baker's solution was to build the new wall bound to the old by iron tie rods, but separated from it by 6 inches (15 cm.) of loose stones. Then later liquid cement would be pumped into the gap to bind the two parts into one whole. The new masonry was laid on top of, and on the front face of the old dam. The new dam, inaugurated at the end of 1912, was 16 feet (4·9 m.) higher, but the new reservoir was 23 feet (7 m.) deeper, holding two and a half times as much water. By 1929 it was necessary to heighten the dam again to provide still more water. This time the old and new masonry were permanently separated by stainless steel sheets. Twenty-nine feet (9 m.) was added to the top of the dam and its thickness was correspondingly increased. The dam had to be made 160 yards (146 m.) longer to accommodate the enlarged reservoir which now stretched back for 184 miles (296 km.). The work was completed in 1933.

One drawback of the Aswan Dam was that it only held enough water for one year's use, so it could not compensate for a low flood. The new Aswan High Dam has now been built and this gives over-year storage as well as hydro-electricity. It is a stratified sand and rockfill dam, a huge inert man-made hill, 350 feet (107 m.) high, half a mile (800 m.) wide and $2\frac{1}{2}$ miles (4 km.) long. Its core is made of clay and it has a grouted cement curtain going down 300–400 feet (90–120 m.) through sand, gravel and boulders, to rock preventing water finding a way under the dam. A grouted cement curtain is made by drilling holes deep in the ground and pumping in liquid cement under pressure. The sand base was compacted to near solidity by vibrators. The lake, when it is full, is as long as England and impounds twenty-six times as much water as the old Aswan Dam, which is 4 miles (6·4 km.) to the north and still used, but only as a regulator of water into the irrigation canals. About 100,000 people had to be rehoused because their homes and villages and towns were submerged. Some of them were retreating for the fourth time. The dam was originally designed by Hochtief-Dortmund of West Germany in 1953 and it was built to these plans with small changes by Russian and Egyptian engineers. It is the largest rockfill dam in the world and is sixteen times the volume of the Great Pyramid.

The hydro-electric station is at the end of six tunnels in a channel in the east bank. The tunnels and channel were built first to take the diverted river while the dam was made, safe and dry behind coffer dams which were eventually incorporated into the finished dam. Work started on 6 January 1960 when President Nasser set off dynamite to shift 20,000 tons of granite in the diversion channel, but not much else was done that year. The Russians and Egyptians had to learn to work together first. The diversion channel is 1,148 yards (1,050 m.) long, into six tunnels 312 yards (285 m.) long and

then another 547 yards (500 m.) of channel back to the lower Nile, all cut
out of solid granite. Twelve million cubic yards of rock were removed and
this was the most difficult part of the whole scheme. In 1963 there were
30,000 men at work twenty-four hours a day in temperatures rising to 135 °F
(57°C) in the shade. In 1963 work began on building the hydro-electric
power station on a concrete platform at the downstream ends of the tunnels.
After the Nile was diverted in 1964 all that remained was to build the high
dam itself. The rock was brought in by an endless convoy of dumper trucks
and barges, and the sand was pumped in as a suspension in water which
drained away leaving the sand *in situ*. There are no official figures but the
Arab Contractors reckoned that about fifty men a year were killed. At least
eleven men died in a rockfall on the cliff above the power station, and six
were killed by explosives that had not detonated completely. A Nubian
worker was standing on rock in a barge and he died when the rock was
prematurely dumped through bottom opening doors. The work was virtually finished in 1970 and the lake began to rise, threatening many ancient
Egyptian monuments.

The most famous of them were the Temples of Abu Simbel, on the east
bank of the Nile near the Egyptian/Sudan frontier, which were the temples
of Rameses II and his queen Nefertiti. They were carved out of the solid
sandstone cliff around 1230 B.C. The hall of the main temple went 200 feet
(61 m.) back into the cliff, the roof being supported by eight square pillars.
The entrance was guarded by four colossi of Rameses II, 65 feet (20 m.) tall
on a façade 119 feet (36 m.) broad by over a 100 feet (30 m.) high. There
were several adventurous civil engineering schemes put forward to avoid
their being submerged. It was proposed to build a dam around them but this
would have been expensive, and aesthetically rather pointless. Professor
Gazzola, Director of Fine Art at Verona, asked a firm of Italian engineers
to work out how to cut the temples bodily out of the cliff and lift them
beyond the reach of the water. The block of the Great Temple would have
weighed 250,000 tons (254 million kg.) in its box of reinforced concrete, and
250 hydraulic jacks, electronically interlocked to stay horizontal, would have
lifted it 200 feet (61 m.). It was quite a feasible scheme and cheaper than
building a dam to protect the temples. Cheaper still was the idea of William
MacQuitty, a Belfast film producer. This was to surround each temple with a
thin and cheap membrane of reinforced concrete, filled on the temple side
with purified water to equalize the pressure on the membrane. People could
view the temples from transparent shafts and corridors, or while eating a
meal in the underwater restaurant. Albert Jaquot, member of the French
Academy of Sciences, proposed to build floating foundations under each

Façade of the Temple of Abu Simbel, carved out of the solid sandstone cliff around 1230 B.C. and due for inundation by the rising waters behind the Aswan High Dam. UNESCO raised money by world-wide appeals and saved the Temple from this fate by cutting up the cliff and reassembling it in a safe position.

temple block and let them float up with rising water. What actually was done was to cut the temples up into small blocks with a serrated wire, like a cheese cutter, and reassemble them using a colour-matched cement.

It was estimated that the Aswan Dam and its associated works, excluding Abu Simbel, would cost £415 million and that this would be paid for in two years by the increase of national income. The hydro-electric station was designed to produce 2,200 megawatts, more than four times the power that Egypt already had.

It was not necessary to build a dam for the world's first major hydro-electric scheme because in effect, nature had provided one. This was at Niagara Falls on the United States–Canada border, and it became operative in 1895. There are two falls really, separated by Goat Island: the American Fall has a drop of 168 feet (51 m.) over a width of 1,000 feet (305 m.) and the Horshoe Fall in Canada is 1,950 feet (595 m.) wide on a curve with a drop of 158 feet (48 m.). About 400,000 tons (400 million kg.) of water fall every minute. There had been many attempts to utilize this energy, but nothing could be done on a major scale until the coming of electricity.

The Niagara Falls Power Company was formed in 1886 with Thomas Evershed, who conceived the basic idea as engineer, but professional engineers from all over the world were consulted over the details. A canal,

Pneumatic drills being used in the construction of the exhaust tunnel at Niagara. The spoil was taken out by a railway suspended from the roof to keep it out of the way—the same idea was later used under the Mersey.

250 feet (76 m.) wide at the mouth, was constructed from the Upper Niagara River 1,500 feet (457 m.) into the United States. A pit was hewn in the solid rock, 463 feet (141 m.) long, 180 feet (55 m.) deep and 20 feet (6 m.) wide, for the turbine hall at the end of the canal. A tunnel 7,000 feet (2,134 m.) long takes the water from the turbines at the bottom of the pit to an outlet in the gorge below the falls. A thousand men worked continuously for three years to make this tunnel. They removed 300,000 tons (300 million kg.) of rock and placed sixteen million bricks in its lining. There were ten turbines of 5,000 hp each in the first station. They used about 6,000 tons (6 million kg.) of water per minute.

The Toronto Power Company employed a similar set-up on the Canadian side. Their tailwater tunnel is 1,400 feet (426 m.) long and emerges right behind the Horseshoe Fall. It is 33 feet (10 m.) in diameter, again through solid rock and lined with brick, except at the exit where concrete rings are used. These rings are designed to drop out as erosion takes the apex of the waterfall back 2 feet (60 cm.) every year. The tunnel was driven from a shaft near the crest of the fall, towards the turbine pit. It was decided to open up the discharge end of the tunnel and dump the spoil out there depending upon the rush of water to take it away. A charge was fired to remove the last curtain of rock, but water poured into the tunnel, which it was not supposed to do as the waterfall is well clear of the cliff, because it was deflecting off a pile of rock. Roped together and working through the spray and water, the miners drilled holes in the boulders and filled them with dynamite. The charges were fired by electricity. The offending rocks were blown into the whirlpool, and the water drained out of the tunnel.

In 1931 when no dam in the world was higher than 400 feet (122 m.), work began on the Boulder Dam. It was to be 727 feet (222 m.) tall. The Colorado was an unpredictable and uncontrollable river. It could turn from a gentle stream to a torrent in a few hours and it was a very rare year when its floods did no damage. This state of affairs could not go on, so the United States Bureau of Reclamation planned a series of dams to control and use the river for hydro-electricity, irrigation and town water supply, the most significant being the Boulder Dam. In its journey from the Rocky Mountains to the Gulf of California, the Colorado River has carved itself 283 miles (455 km.) of gorges; some are 6,000 feet (1,830 m.) deep. Most of these were examined in looking for a likely site for the new dam. Some were only accessible from narrow paths; others could only be surveyed from boats. Two showed themselves to be most suitable: Boulder Canyon and Black Canyon. Test holes were drilled up to 200 feet (61 m.) to sample the rock and Black Canyon was chosen, but the dam is called the Boulder Dam. Sometimes it is known as the Hoover Dam after Herbert Hoover, chairman of the commission charged with the work. The site is 100 miles (160 km.) downstream from the Grand Canyon.

The first job was to build four diversion tunnels, two in each side of the canyon. They are of horseshoe cross-section, 56 feet (17 m.) wide and 42 feet (12·8 m.) high, and their lengths vary from 3,600 to 4,200 feet (1,097–1,280 m.). The rock was hard and needed no shoring, and there was no trouble with underground water. Driving the tunnels became a routine: a carriage was put up to the face, thirty holes were pneumatically drilled and charged, the carriage was withdrawn, the charges fired, and electric shovels moved up to clear the broken rock. The cycle took about ten hours to perform and the tunnel advanced at least 15 feet (4·5 m.) each time. They were finished in less than two years, including a 3-foot (1-m.) thick concrete lining. Sloping shafts into the outer pair of tunnels were made at the same time to be used later for the overflows when the lake behind the dam was full: Lake Mead was to be 584 feet (178 m.) deep at the dam and stretch back for 115 miles (185 km.).

The engineers built coffer dams upstream and downstream of the dam site to keep it dry and divert the river into the tunnels. The base of the dam has a maximum width of 660 feet (201 m.) and goes down at its deepest

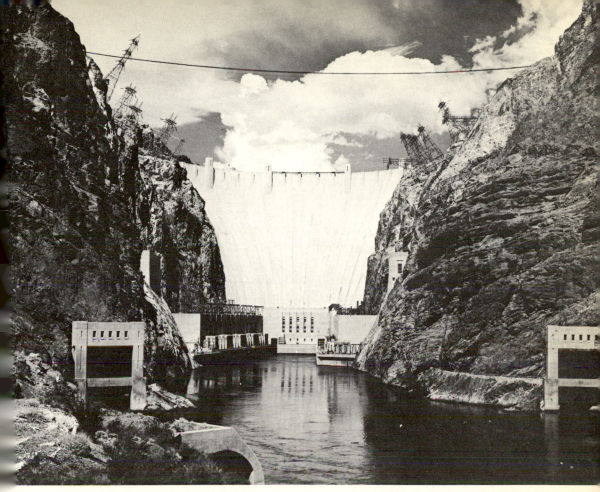

The Boulder Dam on the Colorado River was ready for use in 1936. It contains more than five million tons (5,000 million kg.) of concrete.

139 feet (42 m.) below the river bed. Laying the 5 million tons (5,000 million kg.) of concrete in the restricted space presented some difficulties. An unprecedented problem to be solved was the fact that concrete gives off heat as it sets. Normally this heat is easily lost to the air, but in such a large volume as this the heat would not be able to get away, so the concrete would heat up and expand, and then crack as it cooled. If the concrete were laid layer by layer in the normal manner it would have taken 200 years to construct the Boulder Dam. The answer was to use an immense refrigerator to cool water to near freezing, which was pumped through more than 570 miles (917 km.) of pipes embedded in the concrete as it was poured. The concreting took two years. There are four intake towers on the valley floor upstream, slightly higher than the dam. They are 30-foot (9·1-m.) diameter vertical pipes with openings, supported by a concrete grid, and they take water to the turbines in the power station at the foot of the dam. Fifteen

turbo-generators were installed, each giving 115,000 hp. The dam is arched upstream and the road along its crest is a quarter of a mile (400 m.) long. It was all finished in 1936 when the diversion tunnels were closed by steel shutters.

Modern dams provide irrigation water and electricity; both are essential for the development of areas with low rainfall. Many more dams will be built in the future as the growing world population needs more and more food, and hydro-electricity will be demanded increasingly as fossil fuels run out. But the world should be aware that engineers cannot deliver the goods for ever: eventually the consumer philosophy must change; the population and its demand for power must be stabilized, even reduced.

12

Steel and Concrete

The quite fantastic civil engineering achievements of the twentieth century are due to the availability of large quantities of good quality steel and concrete, and to the engineers' knowledge of how to use them to the best effect. The tensile strength of steel in their skeletal framework makes sky-scrapers possible. The compressive strength of concrete was used to pro-duce the 1,000-foot (305-m.) arch of the Gladesville Bridge over the western arm of Sydney harbour. The two materials are used together in reinforced concrete to make beams and slabs which are cheaper but just as strong as steel alone. The full potential of reinforced concrete was realized by Robert Maillart in his beautiful stiffened arch bridges. Eugène Freyssinet went a step further to create prestressed concrete which has such properties that it may be regarded as a new material.

Steel is an alloy, or solution, of carbon in iron. Other metals may be added as well to give different alloy steels. Cast iron contains much more carbon which makes it brittle but very strong in compression. Wrought iron is pure iron and it can take tensile loads as well which makes it malleable and duc-tile. However, steel is stronger than them both. Large quantities of cast iron were available after 1709 when Abraham Darby successfully smelted iron ore with coke, which made bigger blast furnaces with hotter temperatures possible. Wrought iron was produced in large amounts after 1820 using Henry Cort's puddling process of 1784, just in time to roll railway lines. Like cast iron and wrought iron, steel had been known in the Middle Ages, but it could only be made in small quantities using great skill. In 1856 Henry Bessemer invented his converter which made large quantities of consistent steel. Air was blown through molten cast iron which burned off some of the carbon. The open hearth furnace did the same job, and steel was a moder-ately inexpensive material after 1880.

Nineteenth-century engineers were not slow to use the new materials. The Coalbrookdale Iron Bridge has already been mentioned, so have the iron bridges of Thomas Telford and the steel bridges at St Louis and over the

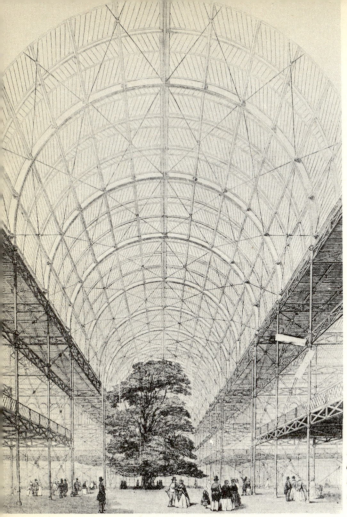

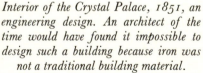

Interior of the Crystal Palace, 1851, an engineering design. An architect of the time would have found it impossible to design such a building because iron was not a traditional building material.

Firth of Forth. Later as the demand arose, iron, and then steel, was used for buildings. Several mills were built in the early nineteenth century using iron as part of their structure. The first large building whose structure was entirely of iron was the Crystal Palace. This was a masterpiece of organization and prefabrication, because it was conceived, designed and erected in nine months. This was possible because Joseph Paxton designed the various members so that they could be mass-produced. There were 3,300 cast iron columns and 250 miles (402 km.) of sash bars all identical. The contractors moved on to the Hyde Park site on 30 July 1850.

Thirty-four miles (55 km.) of drain pipes and concrete foundation sockets for the columns were laid first. A large wooden tripod, called shear legs, was used to erect three columns and two cross-girders in sixteen minutes; then the crew would move on. The first floor columns were dropped into sockets on top of the columns below, and again for the second floor. The hollow columns acted as drain pipes for the internal condensation and the external rain. Then the glaziers moved in. On the roof they worked from a special trolley running in the gutterings of the roof beams. In one week eighty men fixed 18,000 panes of glass. At one stage the glaziers went on strike for more money, but they were all sacked and more men employed in their place. All was finished ready for the opening on 1 May 1851 when a model frigate was to fire a salute of guns. *The Times* was worried that "the concussion will shiver the glass of the roof of the Palace and thousands of ladies will be cut to mincemeat", but all was well. The building was 1,848 feet (563 m.) long and 450 feet (137 m.) wide.

Another pioneer of building with prefabricated standardized parts was James Bogardus working in New York at the same time. His scheme was to hang iron walls on an iron skeleton. The frame took most of the load but some of his walls were loadbearing as well.

The first building to have a true iron skeleton and non-loadbearing walls was the Menier chocolate factory built by Jules Saulnier in 1871–72 at Noisiel-sur-Marne. This four-storeyed building straddled the river on four stone piers to make use of water power for its machines. Its iron frame took all the weight down to the piers, and the walls of hollow brick were true curtains. This French work was apparently unknown to William Jenney who designed and built the Home Insurance Building in Chicago in 1883–85. It had cast iron columns and wrought iron box columns. Wrought iron was also used for the cross-girders and floor beams. Jenney changed his plans during construction and made all the skeleton above the sixth floor of Bessemer steel. There were ten storeys in all. This building cannot be regarded as a mature skyscraper because a small portion of the load was taken by granite piers at the base of the façade, and by the brick party walls at the side; also there was no wind bracing in the frame.

The first building to have all the properties of a skyscraper was the Tacoma Building completed in 1889, also in Chicago. Its iron and steel skeletal frame took all the load and it was riveted together, giving a much greater rigidity against wind forces than the bolted construction of the Home Insurance Building. All the walls of the fourteen-storey Tacoma were curtains and were commenced independently of each other at different levels. There quickly followed many other skyscrapers: the Rand McNally Building was the first to have an all-steel frame in 1890, Jenney's 1891 Fair (now Montgomery Ward) Store is still there, and the twenty-one-storey Masonic Building was the tallest of the period completed in 1892. All these early Chicago skyscrapers were founded on rafts in the underlying clay. Some settlement was anticipated, so building was started on a higher level to allow for this. The skyscraper was invented in Chicago, but New York was not far 157

The Fair, now Montgomery Ward, Store in Chicago under construction in 1891.

behind and there they built much higher, because space was limited.

The first skeletal building in New York was the Tower Building in lower Broadway built by Bradford Gilbert in 1889. The bottom seven storeys had an iron frame but the top five were of traditional masonry. Gilbert had to occupy an office on the top floor of this curious pile to reassure would-be tenants. It was 125 feet (39 m.) tall. Also on Broadway is the Manhattan Life Insurance Building, completed in 1894 and 347 feet (106 m.) high. This was the first building in which compressed air was used to sink pneumatic caissons for the foundations. They go down through 50 feet (15 m.) of mud and gravel to solid rock. This building was also the first to have extensive wind bracing as part of its frame, because the squalls in New York can give wind pressures of 30 pounds per square foot (151 kg. per square m.). In 1913 came the Woolworth Building which took two years to build. For fifteen years, apart from the Eiffel Tower, it was the tallest building in the world with a height of 760 feet (232 m.) and fifty-five storeys above its three basements. Sixty-nine 19-foot (5·8-m.) diameter pneumatic caissons were sunk

to bedrock at an average depth of 110 feet (33 m.) and filled with 70,000 tons
71 million kg.) of concrete for the foundations which support the 223,000-ton
(226,568 million-kg.) building. All the steelwork is encased in concrete to
resist possible fire damage and corrosion. A 3-pound (1·35-kg.) bar was
accidentally dropped from the top during construction and it passed clean
through a tramcar and a suitcase.

At the time the French still claimed that their Eiffel Tower was the tallest
building in the world, while the supporters of the Woolworth Building said
that the Eiffel Tower was not a building. Whatever it was, in 1889 it was a
pioneer piece of engineering, erected said the *Scientific American* "without
error, without accident and without delay". Gustave Eiffel (1832–1923)
was educated at the Ecole Centrale des Arts et Manufactures in Paris from
1852–55. He was employed at first by several railway companies as a civil
engineer until 1867, when he set up his own office as a consulting engineer.

Men at work on the Eiffel Tower in 1889.

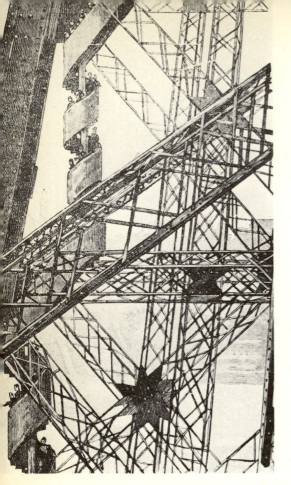

Descending the Eiffel Tower by the stairs, an occupation of 40 minutes.

He was responsible for many civil engineering works all over Europe, and in New York he designed the wrought iron skeleton for the Statue of Liberty in 1885. One of his many notable structures is the Douro Bridge carrying the railway line between Oporto and Lisbon 250 feet (76 m.) above the River Douro. The central crescent-shaped arch has a span of 525 feet (160 m.). It was rather conservative of Eiffel to prefer iron to steel for its main structure, because when work started in 1876, Eads already had successfully completed his three steel arches (page 101) over the Mississippi.

Eiffel used wrought iron, too, for his famous tower. His proposal to erect it for the 1889 Universal Exhibition to be held in Paris was accepted in 1886 and immediately forty calculators and draughtsmen, under Eiffel's direction, set about the task of producing drawings for each one of the 12,000 different bits of metal. These plans were so meticulous that all the factory-made parts fitted on site and none had to be sent back. Meanwhile, in January 1887, work could begin on the foundations. Forty-eight thousand cubic metres of earth were taken out, and 14,000 cubic metres of masonry laid on 2-metre thick layers of concrete as bases for the four piers on which the tower stands. The area of each pier is 15 metres square. In June the iron erectors moved in and by November they had gone up to a height of 30 metres using ordinary shear legs. After that they used twelve free-standing frames of pyramidal form to support four decks, one for each leg of the tower, to reach the first platform and on to the second platform by the same method, but with fewer frames and decks. There was a crane on each deck. The third platform was reached by two cranes on vertical pillars, later intended to be lift shafts. The cranes lifted themselves in 9-metre steps as the tower grew, an operation which took two days. In January 1889, with work nearly completed, there was a minor scare: the tower was not vertical and the second floor was sloping. Accurate surveys and measurements showed this to be an optical illusion. The tower was finished after twenty-six months at the end of March 1889, except for the installation of the lifts. It therefore took the twelve dignitaries who unfurled the tricolour at the top on 31 March forty minutes to get up there. The tower is 984 feet (300 m.) high and its lightning conductor takes it above the 1,000-foot (305 m.) mark.

The dispute about which was the highest building was resolved in 1931 by the construction of the Empire State Building, which is 1,230 feet (375 m.) high to the top of its 200-foot (61 m.) observation tower and airship mooring mast. In the building proper there are eighty-five storeys. It rises on the site area of 425 feet (129 m.) by 200 feet (61 m.) for five storeys; then there is a set-back or small roof terrace. There are further set-backs as the building rises. Owners of adjacent property were worried about the weight of the huge building disturbing their foundations. They appeared to be satisfied with the assurance that the weight of material excavated for its foundations and basements was equal to three-quarters of the weight of the building. It is founded on rock 50 feet (15 m.) below street level and took 3,400 men only fifteen months to build. This was made possible by the overall use of pre-fabricated standardized parts. For example: there are over 3,000 lift gates, all exactly the same on the fifty-eight passenger lifts. The recently completed Sears Tower, Chicago is now the tallest building in the world. Its 109 storeys reach a height of 1,454 feet (442 m.).

The steel skeleton of the Empire State Building was riveted together; today, the members would be welded. Electric arc welding was invented by Auguste de Meritens in 1881 in France, but it was not used on buildings until 1920. A long period of experiment was needed to sort out the local stresses in a frame which are caused by the heat of the weld, and to discover whether the weld had taken all over the contacting surfaces as it should do. In 1920 T. L. McBean used the method to build a factory for the Electric

Welding Company of America. The Westinghouse Electric and Manufacturing Company pushed the process and by 1932 had erected twenty-six such buildings, including the eleven-storey Central Engineering Laboratory in Pittsburgh in 1930.

The other subject of this chapter is made of cement, water, sand and stones. The cement and water react chemically to bind the inert sand and stones. Cement is essentially quicklime which is made by heating limestone. There have been limestone-burning kilns since before 2000 B.C. The Romans discovered that if the lime is mixed with a small quantity of a volcanic ash, called pozzuolana, the cement will set under water. The art of making this hydraulic cement was lost during the Middle Ages and rediscovered around the sixteenth century when various natural supplies of hydraulic lime were discovered. However they were not plentiful and an artificial hydraulic cement was needed. John Smeaton, when building the Eddystone lighthouse (page 42), discovered that a certain proportion of clay was the active ingredient in hydraulic lime, and made a satisfactory cement with hydraulic properties. At the beginning of the nineteenth century Louis Vicat in France, and Joseph Aspdin in Britain, were commercially manufacturing artificial hydraulic cement by burning a carefully proportioned mixture of pure limestone (chalk) and clay. Before his patent of 1824, Aspdin used to ostentatiously carry brightly coloured, but useless, salts into his kiln to deceive his rivals. He called his product "Portland Cement" because when it set, it closely resembled Portland stone. In 1844 Isaac Johnson discovered the modern high temperature process of making Portland Cement, which is to heat the ingredients until they begin to fuse together. The resultant clinker is then finely ground. Hydraulic cement sets with a complex series of chemical reactions between three oxides: of aluminium and silicon (in the clay) and calcium (in the lime). By 1880 concrete was in common use, en masse or as precast blocks, for a whole variety of masonry style uses. The steam-driven concrete mixer had also been developed by that time.

Gladesville Bridge of 1964 is the longest concrete arch in the world. It spans the River Parramatta which forms the western arm of Sydney Harbour. The bridge is 1,900 feet (578 m.) long overall and the arch is 1,000 feet (305 m.) with a rise of 142 feet (43 m.). The four parallel hollow ribs are made of precast concrete boxes or voussoirs, 20 feet (6 m.) wide at the springings to 14 feet (4·3 m.) thick at the crown of the arch. The 72-foot (23-m.) wide road consists of reinforced concrete slabs supported by 2-foot (60-cm.) thick prestressed concrete columns on the arch. The same steel

Gladesville Bridge, Sydney, completed in 1964, has the longest concrete arch in the world with a span of 1,000 feet (305 m.).

centring was used for each arch rib in turn, supported on piles in the river bed, but with a 200-foot (61-m.) wide navigation channel left open. The 50-ton (5,000-kg.) voussoirs were lifted to the centre of the arch from barges, placed on trolleys there, and wheeled into position. Concrete was poured *in situ* to make 3-inch (7·6-cm.) joints between them and to make all the top flanges of the ribs into one. Then the arch was jacked off its centring.

The leading engineers who gathered the knowledge that made the Gladesville Bridge possible were François Hennebique (1842–1921), followed by Robert Maillart (1872–1940) and Eugène Freyssinet (1879–1962). These men were the pioneers and the masters of concrete. Their basic realization was that concrete is a material in its own right. Two particular properties, for example, are creep and shrinkage. After concrete has set, it hardens or cures for anything up to a year, and during this time it shrinks. Creep is the initial inelastic deformation produced by a load while the concrete hardens. Creep and shrinkage caused builders in concrete considerable trouble with cracking; they therefore designed unnecessarily massive, and as a result ugly, constructions to allow for this. Concrete is not a cheap substitute for masonry which was how it had been regarded: in 1889, the first reinforced concrete bridge in the United States had its surface scored to look like cut stones and under the arch there were concrete stalactites. This was the Alvord Lake Bridge in a San Francisco park, a flattened vault with a 20-foot (6-m.) span.

Reinforced concrete is not a new idea; it can even be traced back to Greek times, but it became useful as we know it in the mid-nineteenth century. Two earlier experimenters were François Coignet and Joseph Monier. Monier started in 1867 to make concrete garden pots reinforced with wire mesh, and shortly afterwards he and Coignet both promoted its use for structural elements: beams, floors, pipes, bridges, etc. Concrete, like cast iron and masonry, has virtually no tensile strength. Therefore it breaks very easily if it is used as a beam, because in bending there are tensile strains on the outside curve of the bend. Thus if steel bars are embedded in the concrete where the tension will be, the steel bars will take the tension and the result is a very strong and cheap beam.

One man who emerged ahead of his contemporaries was François Hennebique. In the period 1880 to 1900 he erected over a hundred reinforced concrete bridges in Europe. A notable one is the Pont de Châtellerault over the River Vienne in France, built in 1898–99. It has three arch spans of 131, 164 and 131 feet (40, 49 and 40 m.). The arches have four ribs and a rise of one-tenth of the span. The flat road is supported by spandrel columns. The centring for the arch was supported from the river bed and all the concrete was poured in two and a half months, before the river rose in the late summer. Hennebique's longest span was the 328-foot (100-m.) arch of the Risorgimento Bridge over the Tiber in Rome. It was opened in 1911 and tested by 1,100 soldiers marching over it at the double, followed by three 15-ton (15,000 kg) road rollers running abreast. Such tests as these are only to gain public confidence. No engineer would send 1,100 soldiers and three steam rollers over a bridge that was likely to collapse as a result.

Initially, Hennebique was interested in reinforced concrete for buildings, but there was no demand for such methods in Europe, so, as with the steel skeletal building, it was in America that reinforced concrete was first used for multi-storey buildings. The chief pioneer was Ernest Ransome based in San Francisco, where he started his commercial career as a concrete block maker in 1868. By 1904 he had been responsible for several factories and other buildings. The Alvord Lake Bridge, mentioned above, was his too. He used beds of bars for his reinforcement. The first reinforced concrete skyscraper was Albert Kahn's Brown-Lipe Chaplin Factory in Syracuse, New York, built in 1908. This style of building became very popular after 1910 and is still used. Around 1930 Robert Maillart introduced his beamless floor slabs on mushroom columns. He had been experimenting since 1900

The Brown-Lipe Chaplin Factory, New York, 1908, was the first reinforced concrete building and set a style that became popular.

and had come to the conclusion that, just as beams and girders are natural for steelwork, slabs were the better way to use reinforced concrete. Maillart thoroughly understood concrete and thus produced some very beautiful bridges. They were built between the two world wars but have a very modern appearance. This would suggest that there has been no advance in the use of reinforced concrete since about 1930.

Robert Maillart was born in Berne and went to the Zurich Polytechnic where he qualified as a structural engineer in 1894. He did various jobs for railway companies and then became Assistant Engineer to the Highways Department of Zurich City Council. In 1902 he set up on his own as a contractor, in which capacity he came into close contact with Hennebique who was the consulting engineer for many of his contracts. He went to Russia in 1912 but returned penniless after the war and revolution, and opened his office in Geneva.

His first, and prototype, bridge was over the River Inn at Zuoz, Switzerland, in 1901. Like Hennebique's Risorgimento Bridge, it was a three-hinged arch made from reinforced concrete slabs formed into box cells as wide as the arch. The road slab and the arch slab were thus rigidly joined together and gained strength from each other. A normal arch, like Sydney Harbour

165

Salginatobel Bridge, the longest of Robert Maillart's spans at 293 feet (89 m.).

Bridge, has two hinges: at the abutments. Maillart added a third hinge in all his bridges in the centre. This made it easier for his bridges to accommodate the effects of creep and shrinkage. In 1905 he built a 167-foot (50·8-m.) arch over the Rhine at Tavanasa, Switzerland. It was destroyed by a landslide in 1927. In this one he missed out some of the side slabs so that the spandrels were open, but the road and arch still acted together. This bridge was beginning to take on the "Maillart" shape. All the side slabs were missed out at the Schwandbach Bridge of 1933 to produce the purest of his designs. It is a proper stiffened three-hinged arch, and carries the road on a curve linking the sides of a deep ravine. The inner edge of the arch slab, which is only 8 inches (20 cm.) thick, follows the curve of the road, while the outer edge is straight. Its span is 123 feet (37·5 m.). The longest span that Robert Maillart designed is the 293 feet (89·3 m.) of the Salginatobel Bridge. Parts of the side slabs, or spandrel walls, were left in so that the box section is still retained at the centre. R. Coray, the contractor's engineer, must have had some trouble getting his materials to the mountainous site near Schiers, Switzerland, and erecting the wooden centring over the steep gorge.

The third of this trio of concrete masters was Eugène Freyssinet. He was born in 1879 to a French peasant family and was a student at the famous Ecole des Ponts et Chaussées. He built many concrete bridges in his long life including the three arches of 567 feet (173 m.) on the Pont Albert Louppe Bridge near Brest Harbour in 1930, and more recently, the bridges on the Caracas Autopista in Venezuela with spans of around 500 feet (152 m.) in very difficult mountainous country. He was consulted on the design for Gladesville Bridge by its designers G. A. Maunsell and Partners, but he did not live to see it built. He died in 1962. Freyssinet's great contribution to concrete technology was the art and science of prestressing. He had the idea in 1904, but had to wait until the advent of high tensile steel before it could work to its full effect. There was rapid development in this field after World War II. The idea is to put tension into the reinforcing bars before the concrete sets. This tension is released after the concrete sets and it prestresses the concrete by compressing it in the region where there would be tension when the normal working load is applied. If the compressive prestress is greater than the tension from the load, the concrete will be subject to compression only. This suits the concrete well because it is weak under tension and very strong in compression. This means that very flat arches can be made; so flat in fact, that they look like beams but they are not, because the concrete is all in compression still due to the initial prestress put into it. A prestressed concrete beam requires seventy per cent less

Esbly Bridge over the River Marne, completed in 1950, has a very flat arch of 242 feet (74 m.) made of prestressed concrete.

A unit being lifted into place on the Esbly Bridge. This and four other bridges over the Marne are identical.

steel and about forty per cent less concrete than a reinforced concrete beam of the same loadbearing capacity. Another way to prestress concrete is by tensioning wires threaded through ducts left in the precast concrete. This way several precast sections of a bridge can all be prestressed as one unit, after their erection.

The classic example of Freyssinet's work are the five bridges over the River Marne at Esbly, Annet, Tribardos, Changis and Ussey. All are identical flat arches with spans of 242 feet (73·7 m.) and they were erected between 1948 and 1950. The arches are made of prestressed I-section voussoirs, mass-produced in a factory. The flanges were cast first with wires connecting them. Then the 4-inch (10-cm.) thick webs were cast with the flanges jacked apart putting a tension in the wires. When the jacks were removed, the voissoirs were prestressed vertically. This stopped the web getting diagonal cracks due to the shear forces of the load after it was erected. The voussoirs were stored before use so that they would be cured properly before the main prestressing was applied by wires in ducts. The voussoirs were joined together in sets of ten and prestressed. Two such groups were joined side by side to make a manageable unit. Three units side by side make up the six ribs of all five bridges. Two masts were put up at each end of the bridge site. Tackle from both of them could lift a unit into any position and cables from the mast also held up the erected units until they were all in place.

It is likely that steel and concrete used together have not yet achieved their full potential, particularly in the field of building construction. The advent of the vertical town is not far away; its beginnings can be seen in the Hancock Centre in Chicago. This multi-purpose building has car parking and stores on lower floors, offices at the middle level and accommodation in the upper storeys. The 100-storey building rises to 1,106 feet (337 m.).

13

Wire Suspension Bridges

A suspension bridge could be regarded as an inverted arch. Except for the road deck, which is subjected to small and local bending moments and is therefore stiffened accordingly, all the stresses are tensile. Materials such as steel can take a bigger load in tension than they can in compression because a strut will fail due to buckling before it will fail under compression alone. Another point which makes the suspension bridge the most economical type for long spans, is its flexibility. It gets extra strength from this by moving in order to distribute the stresses placed upon it. But this has been the downfall of many a suspension bridge: if it is too flexible it can be set in motion by only a moderate wind, the motion comprising undulating and torsional waves.

The Menai Straits Bridge (page 63) was opened in January 1826. Storms in February started the bridge undulating and twenty suspenders were broken and many more bent. On 1 March, waves on the bridge caused the horses of the Chester Mail to fall down. This led to bracing chains being fitted with stronger suspenders and road bars. In a storm of 1839 part of the deck fell into the Straits.

Telford had experimented with wire cables while he was designing his Menai Bridge, but at that time and for that span, it was safer to use eyebar chains as he did. When iron is cold drawn into wire it has a higher tensile strength; thus a cable made of many wires is stronger than a bar with the same total cross-sectional area. A wire cable is also better than a chain for a suspension bridge because it is more perfectly flexible.

The fundamentals of the wire suspension bridge were developed in France between 1825 and 1830 by Marc Seguin, who constructed several with spans up to 300 feet (91 m.) stiffened by truss girders, and by M. Vicat, who devised a method of spinning the cables *in situ* wire by wire. In 1834, Joseph Chaley, an associate of Seguin, finished such a bridge over the Sarine Valley at Fribourg, Switzerland. Its 870-foot (265 m.) span was supported by four 5½-inch (14-cm.) diameter wire cables. These ideas were taken to the United

Street Bridge, Philadelphia, built by Charles Ellet over the Schuylkill River in 1841–42. Ellet's practice of keeping the strands separate can be discerned.

States by Charles Ellet and John Roebling to reach their full development in the Brooklyn Bridge.

Charles Ellet (1810–62) was born in Pennsylvania and went to the Ecole Polytechnic, Paris in 1830 at his own expense to finish his engineering education. He toured England and Germany and returned to America in 1832 where he designed and built several wire suspension bridges. His longest was over the Ohio River at Wheeling. Its 1,010-foot (307·8-m.) span was 97 feet (30 m.) clear of the river, but it was wrecked after five years by a storm in 1854. One of the troubles was that Ellet laid his strands of wires side by side, and connected them by bars of iron from which hung the suspenders for the roadway. This meant that the strands were exposed to wear and corrosion, and they acted separately. When John Roebling rebuilt the Wheeling Bridge, he squeezed the strands together to make a round cable bound with wire. The strands now acted together as a unit. The suspenders were hung from iron clamps round the cable.

Ellet planned a suspension bridge over the Niagara Gorge, 2 miles (3·2 km.) below the falls. It was to be 800 feet (244 m.) long for two roadways, two footpaths and one railway track. In 1848 he constructed a light service suspension bridge 770 feet (235 m.) long and $9\frac{1}{2}$ feet (3 m.) wide. He was so pleased with it that before the handrails were erected he rode across on

*Roebling's Niagara Suspension Bridge, 1855, showing the stiffening stays radiating from the
tops of the towers which he used again on his Brooklyn Bridge.*

horseback 250 feet (76 m.) above the rapids. Then he had a quarrel with the
directors of the bridge company and withdrew from the contract. The
footbridge stayed in use until 1854–55 when Roebling built a road and
railway suspension bridge.

John Augustus Roebling (1806-69) was born in Germany and graduated
as a civil engineer from the Royal Polytechnic School, Berlin in 1826. He
became involved unsuccessfully in revolutionary politics and emigrated to
the United States in 1831. After an eleven-week journey in which they were
chased by pirates and storms, his ship docked at Philadelphia. Roebling
tried farming in west Pennsylvania and then worked for various canal companies. He built five aqueducts using the suspension principle; the biggest
took the Pennsylvania Canal across the Allegheny River in 1844 on seven
spans of 162 feet (49·3 m.) each, supported by two 7-inch (18 cm.) diameter
wire cables. In connection with his canal work Roebling established his own
wire factory. After he completed the Niagara Bridge in 1855, Roebling
started on a suspension bridge over the Ohio River at Cincinnati and this
proved to be very difficult. The winters were cold with ice and floods to
contend with, and work had to stop from 1859 to 1863 when the money
ran out. The Civil War did not help either. By 1865 the masonry was completed and the towers stood 230 feet (70 m.) high and 1,057 feet (322 m.)

apart. The 5,180 wires were spun, and bound into seven strands for each cable. Then the cables were wrapped in wire and the suspender bands shrunk on by putting them on hot. The bridge was opened on 1 December 1866 and during the next two days 166,000 people walked across.

These bridges would deserve a much more detailed description if it were not for the mighty Brooklyn Bridge, the longest span in the world until the Forth Railway Bridge. It was the crowning achievement of John Roebling's career and cost his life and the health of his son. In 1867 he was appointed engineer to design and build the bridge from Brooklyn to New York over the East River. The centre span was 1,595$\frac{1}{2}$ feet (486·3 m.) with two side spans of 930 feet (283·5 m.) which, including the approach arches, made a total length of 5,989 feet (1,825·5 m.) It carried road and light railway traffic until 1951 when the railway was removed and now there is a six-lane highway 85 feet (25·9 m.) wide.

When Roebling announced his scheme there was a lot of fuss from the ignorant and the prejudiced. After all it was to be fifty per cent longer than the difficult Cincinnati Bridge, it was intended to carry a much bigger load, and steel wires were to be used for the first time. It took Roebling two years to quieten his critics before he got permission to build in June 1869. In the same month he was standing on a wharf surveying to fix the position of the Brooklyn pier, when a boat came in and crushed his foot. His toes were amputated but tetanus developed and he was dead within three weeks. All the plans for the bridge had been completed, but people thought that now the inspiration had gone the bridge could not be built. But John Roebling had an able deputy in his son Washington.

Washington Roebling (1837–1926) had graduated as a civil engineer from the Rensselaer Polytechnic Institute, the first engineering school in the United States, and had worked with his father on the Cincinnati Bridge and

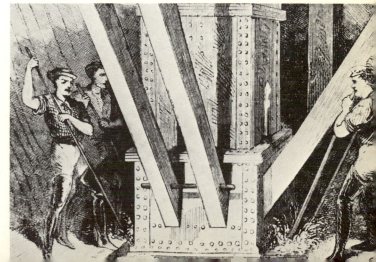

The bottom of the water shaft in the caisson on the Brooklyn Bridge.

in the planning of the Brooklyn Bridge. He now took over and started work on founding the Brooklyn tower 44 feet (13 m.) below the high water mark. A pneumatic caisson was used to get through the mud on the river bed to a solid conglomerate of clay, sand and boulders. (In a pneumatic caisson, air pressure is used to keep the water and mud out of the working chamber at the bottom. There men work, excavating material and the caisson sinks to a solid foundation. See page 102.) The caisson was made of timber and measured 102 feet (31·4 m.) by 168 feet (51·1 m.). The working chamber was divided into six compartments with $9\frac{1}{2}$ feet (2·7 m.) headroom below the 15-foot (4·5 m.) thick roof. The sides were 9 feet (2·7 m.) thick tapering to an iron shod cutting edge. The New York caisson was similar, but the roof of its working chamber was 22 feet (6·7 m.) thick because it was to go to the greater depth of 78 feet (24 m.) through sand to solid rock. The Brooklyn caisson was put into position and men entered the working chamber for the first time in May, 1870. Each caisson had two water shafts which went down below the bottom level of the caisson into a pit. The pit was kept filled with water, which rose up the shaft to balance and seal the air pressure in the working chamber. Excavated material was put into the pit and lifted up the shaft through the water without having to pass through an airlock. On one Sunday morning when no one was in the working chamber, the water in the pit fell below the level of the shaft and the air pressure escaped. A great shower of mud, stones and fog roared up the shaft and a few hundred feet into the air.

These were the days just before electric power, so Roebling had to use candles and gas burners for lighting, which inevitably led to a few fires in the pressurized working chamber of the Brooklyn caisson. The New York caisson had its timber lined with iron plate to prevent this. The last fire was the largest. On 2 December 1870 a candle had been placed on a shelf too near the roof which ignited. The air pressure drove the fire into the interior of the timber and no flames or smoke were visible until the fire had a good hold. Carbon dioxide gas was used, and water and steam applied with a hose, all to no avail. Washington Roebling was in the chamber fighting the fire for seven hours and was brought out unconscious. The fire was still alight so he ordered the caisson to be flooded. After two and a half days the water was pushed out by reapplying the air pressure. All the burned timber was cut out and the space filled with new wood and concrete. These repairs took three months. Then the Brooklyn caisson could go down to foundation level where the working chamber was filled with concrete and the tower completed on top. The New York caisson was sunk in a similar way but it went deeper (78 feet) (24 m.) and there were many cases of the "bends",

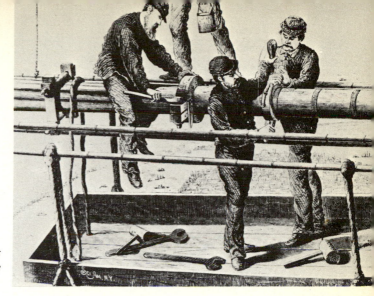

Compacting the strands into cables and binding them with wire on the Brooklyn Bridge.

over a hundred of them serious. Washington Roebling probably spent more time than anyone in the working chambers under pressure. In the spring of 1872 he was brought out partially paralysed and the use of his voice was gone for good. He directed the rest of the operations from a nearby house using a telescope and written messages delivered by his wife.

Both towers were finished by the middle of 1876 to their full height of $271\frac{1}{2}$ feet (82·7 m.) ready for the spinning of the cables. Eighty-five thousand cubic yards of granite and limestone were used in the towers. The cables were spun by pulling a 5-foot (1·52 m.) diameter pulley on a rope along the intended line of the cables. Each time it left the Brooklyn shore it carried a loop of wire, one sixth of an inch (0·44 cm.) in diameter, which was fixed to the anchorages on the New York side on arrival after ten minutes' travelling. The pulley returned empty. When 286 wires were spun they were bound together to form a strand. Some of the strands had extra wires to make up for some deficiencies in the wires already spun. On 14 June 1878 the apparatus for picking up a new strand gave way. That strand shot over the tower and into the water, killing two men on its way. When seven strands had been laid, they were bound together every 10 inches (25 cm.) into a 9-inch (23 cm.) diameter cable. Then a further twelve strands were laid around it to make a 16-inch (40-cm.) diameter cable. There are four cables and it took twenty-six months to spin them using about 1,200 miles (1,900 km.) of wire. There was no solid rock in which to construct the anchorages for the cables so Roebling had to hold down each 23 ton (23,000 kg.) cast iron anchorage plate with 44,000 tons (44 million kg.) of masonry at the end of each side span. The cables go through the towers on saddles free to move on wrought iron rollers so that the towers only have to take a vertical load. The saddle plates are 147 feet (45 m.) above the road. Special machines were used to

176   *The towers of the George Washington Bridge, originally intended to be stone-clad, are 595 feet*
(181 m.) high.

squeeze and completely wrap the cables in galvanized steel wire at the rate of 20 feet (6 m.) a day.

The road platform was made by six parallel steel truss girders supported by suspenders from iron clamps round the cable. The girders are 119 feet (36 m.) above the river at the towers and rise to give a clearance of 135 feet (41 m.) at the centre. Added support and stiffness is gained by stays radiating from just below the saddles in the towers and fixed to the platform at 15-foot (4·6-m.) intervals for 150 feet (46 m.) each side of the towers. Roebling also used these stays on his Niagara Bridge (picture, page 172). Originally the 85-foot (26-m.) wide platform carried two carriageways, two tramtracks and a footpath. Brooklyn Bridge was finished in the spring of 1883 after fourteen years. Washington Roebling lived on, although confined to a chair, running the family business until he died at the age of eighty-nine.

The George Washington Bridge, opened in 1931, has a main span of 3,500 feet (1,067 m.)

The next major advance in suspension bridge design was the George Washington Bridge over the Hudson River. Its towers were designed by O. H. Ammann on the same principle as a skyscraper, that is: a steel frame clad with non-load bearing stone. Thus they could be higher and stronger than traditional masonry towers as used by John Roebling on his Brooklyn Bridge, and they enabled the 3,500-foot (1,067-m.) span to be more than twice that of the Brooklyn Bridge. In the event the stone cladding was not put on the 595-foot (181-m.) towers. The foundations of the New Jersey tower were built inside two coffer dams, bigger and deeper than any previous coffer dams. Each was 108 feet (33 m.) long by 99 feet (30 m.) wide, made of two rows of sheet piling driven to depths of 40–85 feet (12–26 m.). The 8-foot (2·4-m.) gap between the two rows was filled with concrete at the lower level and with ballast on the top. The four suspension cables, each 3 feet (91 cm.) in diameter and made of 26,474 parallel wires were spun, compacted and bound in the air in the way that Roebling and his predecessors had pioneered. The towers and cables were designed to take the weight of a double deck road, but it was only considered necessary in the first instance to build a single deck. It was erected outwards from the towers by the travelling cranes on the deck itself. The George Washington Bridge was opened in October 1931 after four years of work. The second, and lower deck, was added in 1959–62 with 30-foot (9-m.) stiffening trusses between it and the upper deck.

San Francisco does rather well for suspension bridges and it needs them, isolated as it is at the north end of a peninsula enclosing the San Francisco Bay. There are two bridges: the Bay Bridge which goes east to the mainland and already described in Chapter 10, and the famous Golden Gate Bridge linking San Francisco to Marin County in the north across the mouth of the bay.

The Golden Gate Suspension Bridge has a span of 4,200 feet (1,280 m.). Its 90-foot (27·4-m.) wide deck is suspended from two 3-foot (91-cm.) diameter cables on towers 746 feet (227 m.) tall. Its construction was under the direction of J. B. Strauss as chief engineer. His biggest trouble was the foundation for the south tower, 1,125 feet (343 m.) from the shore and exposed to the open sea with a tidal current of seven and a half knots. Strauss planned to sink a steel caisson inside a permanent concrete coffer dam, or fender, to protect it from the sea. His first move, in 1933, was to build a trestle bridge from the shore to the site for easy access. This was supported on piles in holes in the rock of the sea bed. No sooner was it finished, than a steamer crashed through it in fog. He rebuilt it only to have 800 feet (243 m.) carried away by a storm. Again it was replaced. Next the fender had to be built at the end in 65 feet (20 m.) of water. Deep sea divers on this job could only work for short periods at the slack water between tides. They placed six 'bombs', 8 inches (20 cm.) in diameter by 20 inches (50 cm.) long, to blast off 15 feet (4·6 m.) of rough rock from an area 311 feet (95 m.) by 170 feet (52 m.). The shattered rock was removed by dredging. They built the fender in twenty-two sections each 33 feet (10 m.) long. Concrete was poured underwater into steel shutters held in place by a guide frame on the trestle bridge. The sections rose 15 feet (4·6 m.) above the water but eight on the east side were left 40 feet (12 m.) below, so that the 8,000 ton (8 million kg.) steel caisson could be floated in. This was duly done in October 1934, but it was damaged when a heavy swell bashed it hard against the fender. It would take a fortnight to close the fender and in the meantime the caisson would be wrecked.

Strauss decided to change his plans and he had the caisson towed 30 miles (48 km.) out to sea and dumped. The fender was completed and another 29 feet (8·8 m.) of concrete was laid across the base underwater on top of the 36 feet (11 m.) already there. The water was pumped out and the south tower built straight on this base. He had no trouble with the rest of the bridge except that ten men working on scaffolding under the deck were killed when the scaffolding broke away and crashed through the safety net, taking them 240 feet (73 m.) down to the water.

The Golden Gate Bridge was opened in 1937. It took the world record span away from the George Washington Bridge and held it until O. H. Ammann hit back with the Verrazano Narrows Bridge at the entrance to New York Harbour. The Verrazano Narrows Bridge with a 60 foot (18·3 m.) longer span of 4,260 feet (1,298 m.) was opened in 1965. It is a massive conventional design using 150,000 tons (150 million kg.) of steel and there are 142,500 miles (229,325 km.) of 0·196 inch diameter cold-drawn galvan- ized steel wire in its four 3-foot (91 cm.) diameter cables.

Suspension bridge engineers were obviously getting very confident in the 1930's. Their bridges were getting lighter, more slender and cheaper. The Golden Gate Bridge was particularly so. The side trusses giving stiffness to the deck were only 29 feet (11·8 m.) deep and it was estimated that the deck would swing at midspan by 42 feet (13 m.) in a 120 mph (190 kph) wind. This would not harm the bridge. An indication of things to come occurred in 1938 in a 62-mph (100-kph) wind when 2-foot (60-cm.) waves were observed running along the deck. Four thousand seven hundred tons (4,775,000 kg.) of lateral bracing have since been added.

Slenderness went a step further with the now notorious Tacoma Narrows Bridge across Puget Sound in Washington State. Its 2,800-foot (853-m.)

span, third longest in the world at the time, was only 39 feet (11·8 m.) wide and stiffened by 8-foot (2·4-m.) deep plate girders. According to engineering theory at the time the bridge was perfectly safe in any winds up to 120 mph (190 kph), but this did not take into account aerodynamic effects. The Tacoma Narrows Bridge was opened on 1 July 1940. The deck swayed sideways and there were noticeable vertical waves. Hydraulic buffers were installed between the deck and the towers to dampen the motion, and diagonal suspenders were added. On 7 March 1940 the wind was blowing at 42 mph (68 kph). This set the deck heaving in vertical waves up to 30 feet (9 m.). After three hours this motion suddenly changed to a torsional motion with the deck twisting through 90°. Then a series of suspenders broke and 1,000 feet (300 m.) of the central span fell into the water. The motion stopped immediately but then appeared in the side spans. This caused the rest of the main span to fall and finally all motion ceased. Laboratory and wind tunnel tests showed that the deck of a suspension bridge can act like an aeroplane wing, having lift and drag in certain conditions. Two attacks can be made on this trouble: one is to make the deck stiffer and heavier, and the other is to stop the lift by flaps and fins, or equalize the pressure above and below by having openings between the carriageways.

British engineers were particularly interested in this work because definite plans were afoot for suspension bridges over the River Severn and the Firth of Forth. Design work and wind tunnel tests went ahead for both bridges, but the Forth Bridge was given priority of building and was finished in 1964 after six years of construction. Its main span is 3,300 feet (1,006 m.) and there is nothing remarkable in its design. Wind tunnel tests continued on the Severn Bridge design in the light of the experience gained on the Forth and it was found that the side trusses could be reduced. Gilbert Roberts of Freeman Fox and Partners, the chief designer for the bridge, arrived one day to find that the one hundredth scale model of the Severn Bridge had been smashed in the wind tunnel at Bedford through not being fixed down firmly enough. It would take some time to make another model, so Roberts decided to use that time in experimenting on box section road decks with different shaped triangular sidepieces. These proved so successful that the intended traditional deck with open lattice side trusses was abandoned and a revolutionary new design was thus produced, saving twenty per cent on the steel and £800,000 on the cost. The only drawback of the new deck was a small vibration at low wind speeds. Roberts overcame this by making the suspenders incline as inverted Vs instead of the traditional vertical ones thus damping out the vibration.

The towers of the Severn Bridge were also economical in their use of steel.

A section of the deck being raised on the Severn Bridge.

It had been general practice, ever since the Camden Bridge of 1926 in Phila-delphia which was the first, to make the towers of a cellular construction, but the towers on the Severn Bridge are each made of four stiffened steel plates. As a result of these innovations, the Severn Bridge is another major advance in suspension bridge design. The Verrazano Narrows has more than twelve times the dead weight per foot of length, uses twenty-seven times as much steel in its towers and cost about ten times the price. Of course, the Verrazano Narrows Bridge carries three times as much traffic, and has towers of nearly twice the height because its span is thirty-two per cent longer, but nevertheless, the Severn Bridge is by far the more economical design.

The Severn Bridge runs over the Severn Estuary between Aust Cliff and Beachley Peninsula, 8 miles (13 km.) upstream from Avonmouth where the river is a mile (1·6 km.) wide at high tide. The main span is 3,240 feet (987·5 m.). During cable spinning, the 400-foot (122 m.) towers were strained by cables back towards the shores by 2 feet 6 inches (75 cm.) so that they would resume their proper vertical position when the cables and

deck were in place. The deck sections, 104 feet (31·7 m.) wide by 10 feet (3 m.) deep and 60 feet (18·2 m.) long, were fabricated on slipways at Chepstow and launched into the River Wye. As they were boxes they floated and thereby reduced the problems for K. E. Hyatt who was in charge of the superstructure's erection. He had previously experimented with pontoons having outboard motors on each corner, and this led to the design and manufacture of a special "pusher" barge to manoeuvre the floating deck sections into position under the bridge. Hyatt had no difficulties in hoisting the 130-ton (130,000 kg.) deck sections into place, attaching them to the suspenders and welding them together. In fact, the whole contract was remarkably free from troubles, due largely to careful planning and preparation. There were only two fatal accidents in the five years and the boat on permanent stand-by to pick fallen men out of the river was never needed. Labour stoppages amounted to one hour.

John Howard and Company started work on the foundations in May 1961. It was only possible at first to work for two sessions of twenty minutes a day on the Aust (east) pier because of the tidal conditions; the range is over 40 feet (12 m.) with currents of ten knots. The superstructure was commenced in March 1963 and the bridge was opened to traffic on 8 September 1966. The contractor for the superstructure was Associated Bridge Builders, a consortium formed by Sir William Arrol and Company, Cleveland Bridge and Engineering Company, and Dorman Long Limited.

In 1973 the same contractors and consulting engineers began work on what will be the longest span in the world. This is the Humber Suspension Bridge, five miles (8 km.) west of Kingston-upon-Hull, with a main span of 4,626 feet (1,410 m.) and side spans of 910 and 1,723 feet (277 and 525 m.). Its towers will rise 518 feet (158 m.) above the high water mark and will be made of concrete. It is due for completion at the end of 1976.

The suspension bridge has come a long way since the simple rope bridge which can still be seen in the primitive societies of the world. The civil engineer has come a long way, too; his achievements have come in answer to the demands of society and they have advanced that society. The new society thus formed has made fresh calls on the engineers' talents and he has never failed to respond so far. Engineering and society work together, pushing each other on, bringing change and progress.

Left: *The Severn Bridge, a view showing the join between the deck and cable.*   183

Suggestions for Further Reading

Atkinson, R. J. C., *Stonehenge* (Hamilton, 1959)

Beaver, Patrick, *The Crystal Palace* (Evelyn, 1970)

Beaver, Patrick, *A History of Tunnelling* (Peter Davies, 1972)

Beckett, D., *Great Buildings of the World: Bridges* (Paul Hamlyn, 1969)

Bill, M., *Robert Maillart* (Pall Mall Press, 1969)

Coleman, T., *The Railway Navvies* (Hutchinson, 1965)

Condit, C. W., *American Building* (University of Chicago Press, 1968)

Edwards, I. E. S., *The Pyramids of Egypt* (Pelican, 1947)

Engelbach, R., *The Problem of the Obelisks* (Humphries, 1923)

Gies, J., *Adventure Underground* (Hale, 1962)

Hammond, R., *Civil Engineering Today* (Oxford University Press, 1960)

Home, G., *London Bridge* (Lane Bodley Head, 1931)

Kirby, R. S., *Engineering in History* (McGraw Hill, 1956)

Majdalany, F., *The Red Rocks of Eddystone* (Longmans Green, 1959)

Overman, M., *Roads, Bridges and Tunnels* (Aldus Books, 1968)

Prebble, J., *The High Girders* (Martin Secker and Warburg, 1956)

Pudney, J., *Suez: De Lessep's Canal* (Dent, 1968)

Rolt, L. T. C., *Isambard Kingdom Brunel* (Longmans Green, 1957)

Rolt, L. T. C., *Thomas Telford* (Longmans Green, 1957)

Rolt, L. T. C., *Great Engineers* (Bell, 1962)

Rolt, L. T. C., *George and Robert Stevenson* (Longmans, 1950)

Schreiber, H., *History of Roads* (Barrie and Rockliffe, 1961)

Shirley-Smith, H., *The World's Great Bridges* (Phoenix, 1964)

Smiles, S., *Lives of the Engineers* (David and Charles, 1968)

Von Hagen, V., *Roads that led to Rome* (Weidenfeld and Nicolson, 1963)

Acknowledgements .

I would like to thank all the people who have helped and encouraged me with this book, in particular my father, Leonard Upton, who did all my photographic work, and my editor, Rosemary Davies.

Acknowledgements are due to the following for permission to reproduce pictures on the pages indicated: Aerofilms Ltd., 19; *Appleton's Cyclopaedia of Applied Mechanics* (1892), 151; Balfour Beatty & Co. Ltd. (a member of the BICC group), 125, 126, 127, 128, 129; Cement and Concrete Association, 139, 140, 166, 167T, 167B; Clark, E., *Britannia and Conway Tubular Bridges* (1850), 93; Clark, Somers and Engelbach, R., *Ancient Egyptian Masonry* (O.U.P., 1930), 11; De Maré, Eric, *Bridges of Britain* (Batsford, 1954), 48; Photo Deutsches Museum, München, 33; Egyptian State Tourist Office, 12; Engelbach, R., *The Problem of the Obelisks* (Humphries, 1923), 17; Freeman Fox and Partners, 181; French Tourist Office, 46; General Motors Corporation, 165; German State Railways, 133; Guildhall Library, City of London, 49, 50, 51B; William Heinemann Ltd., 15, 60, 76, 118, 136; Illustrated London News (1849), 95, (1851), 156, 158, (1869), 79, 80, (1880), 114; Layson, J. F., *Great Engineers* (Walter Scott, 1893), 159, 160; Leicester Museum and Art Gallery, 90; Library of Congress, 144; G. Maunsell and Partners, Melbourne, 163; McGraw Hill Book Company, 171; Mersey Tunnels Joint Committee, Liverpool, 134, 135, 137T, 137B; Perring, J. S., *The Pyramids of Gizeh*, (1839/42), 14; Phillips, P., *Forth Bridge*, (Grant, Edinburgh, 1889), 110T, 110B, 111T, 111B, 112, 113; The Port of New York Authority, 176, 177; Proceedings of the Institution of Civil Engineers (1909/10), 121T, 124; Redpath Dorman Long Ltd., 107, 108; Riou, Edouard, *Magazine l'Illustration* (1869), 77; Robertson, H. R., *Life on the Upper Thames* (1875), 37; Photo Mike Saunders, 182; Schreiber, T., *Atlas of Classical Antiquities* (1895), 10, 31, 34, 35; Photo Science Museum, London, 24, 41, 43, 51T, 55, 63, 83, 85T, 85B, 87, 88, 89, 92, 94, 98, 99, 100, 106, 172; Shirley-Smith, H., *The World's Greatest Bridges* (Phoenix, 1953),

173, 175; Simms, F. W., *Practical Tunnelling* (Technical Press, London, 1896), 121B; Smiles, S., *Lives of the Engineers* (1862), 52, 67, 69T, 69B, 70; Spanish National Tourist Office, 30; UNESCO/Laurenza, 150; Photo Neil Upton, 20, 21, 23, 53, 58, 62T, 62B, 64, 73T, 73B, 96; US Bureau of Reclamation, 153; Williams, Archibald, *Engineering Wonders of the World* (1909), 147; Woodward, C. M., *History of the St. Louis Bridge* (Jones & Co., St. Louis, 1891), 102, 103, 104, 105.

Index

Bold figures refer to illustrations.

187